THE APARTMENT CARPENTER

BY HOWARD FINK

quick fox

New York   London

International Standard Book Number: 0-8256-3053-3
Library of Congress Catalog Card Number: 74-28708
Printed in the United States of America.

In Great Britain: Book Sales Ltd., 78 Newman Street,
London W1, England.

In Canada: Gage Trade Publishing, P. O. Box 5000,
164 Commander Blvd., Agincourt, Ontario M1S 3C7.

Book and cover design by Elliot Epstein
Illustrations by Lily Hou

CONTENTS

1 GETTING STARTED

THE APARTMENT

The apartment carpenter, by definition, works in an apartment. Be it a one-room studio or a seven-room suite, an apartment always presents certain problems: space, time, and money. There is no place to set up that basement workshop filled with power tools, racks of hanging jars, and pegboards; we have to work in spaces that are also our living rooms or kitchens or bedrooms and no one wants to live around an unfinished project that takes a great length of time. As for money, the difficulty is how to build attractive furniture without investing in the numerous power and hand tools that most people think are necessary.

The methods and projects in this book are designed to minimize these problems. With a little planning you can build a lot of sturdy and handsome furniture in a day or two with a set of tools that costs approximately fifty dollars. Many projects call for precut lumber which is easier to handle and store in the smaller work space of an apartment, and takes advantage of the lumberyard's expensive power equipment.

WHERE TO WORK

Any space that allows you to move comfortably may be your workroom. You will need enough room for a table, which will be your workbench, and enough room to move around it.

The worktable should be large and solid, like an old dining-room table. It must have a lip less than an inch thick and projecting at least an inch from the rails that run around the underside of the table. If your table has no lip, you can construct an auxiliary table that C clamps to the rails—those boards which support the table's edge. Making an auxiliary table can be good practice in constructing simple, useful devices for your apartment workshop. The table should be moveable and resistant to the inevitable scratches and gouges of woodworking. If it isn't scratch-resistant, put several thicknesses of newspapers on the table and top with a sheet of plywood, or make an auxiliary table.

A workroom requires both general illumination and task lighting. Overhead lights plus natural light (if you can get it) are sufficient for general illumination. Task lighting should illuminate the subject, such as a line drawn on a piece of wood, fill in shadows caused by overhead lighting, and avoid glare.

Carpentry is messy. There is always sawdust, dripping glue and varnish, and dropped nails. Therefore, it is best to work in a room with a tough and easy-to-wash floor—or else buy a large strong dropcloth. Be extremely careful with stray nails and screws, which can leave long, deep gashes in almost any flooring if they are stepped on. Keep a wastebasket right next to you and immediately throw all waste into it rather than onto the floor. Learning to work clean will save a lot of after-work cleanup.

STORAGE AND HANDLING

All the tools you will need can fit in a dresser drawer in a large, strong box, preferably one with compartments to keep tools separate and safe. Most important is that they be packed with some care so you don't have a tangle of blades and points that you have to fish around in to find what you want. Keep the protective sheaths and wrappers that come with many tools.

Handling and storing lumber is more difficult. The major problem with lumber is getting it into your apartment, which is why so many of the projects here are designed for precut lumber. By having your projects outlined before buying the wood everything can be cut at the lumberyard to the exact sizes needed.

In the apartment workroom, put the lumber out of the way—against the wall or under the worktable or in the nearest hallway. You want it near enough to get when you need it, but far enough so you are not constantly stepping around it as you work. Leftover wood goes wherever it is out of the way; what you save depends on what you are willing to live with.

NOISE

It is very difficult to diminish noise in carpentry. Inexpensive power tools and hammering create loud sounds.

Try to do noisy work during the day and quiet activities, like painting or sanding, at night. Try to avoid letting wood or tools fall to the floor, which can be irritating your neighbors below.

SAFETY

Carpentry is exciting and engrossing work; however it is too easy to get involved in your work and hurt yourself. Safety is largely a habit—training yourself to practice the rules of safety automatically. Learn to work with the following rules in mind and you can prevent a lot of needless pain and trouble.

1. **Keep work area clean.** Cluttered areas and benches invite accidents.

2. **Avoid dangerous environment.** Do not expose power tools near water. Keep work area well lit.

3. **Keep children away.** All visitors should be kept a safe distance from work area.

4. **Store idle tools.** Tools not in use should be stored in dry, high or locked place, out of reach of children.

5. **Don't force tool.** It will do the job better and safer at the rate for which it was designed.

6. **Use right tool.** Don't force small tool or attachment to do the job of a heavy-duty tool.

7. **Wear proper apparel.** Avoid loose clothing or jewelry which can get caught in the moving parts.

8. **Use safety glasses.** These are necessary with most tools. Also use face or dust mask if cutting operation is dusty.

9. **Don't abuse cord.** Never carry tool by cord or yank it to disconnect from receptacle. Keep cord from heat, oil, and sharp edges.

10. **Secure work.** Use clamps or a vise to hold work, for safety and to free hands to operate tool.

11. **Don't overreach.** Keep proper footing and balance at all times.

12. **Maintain tools with care.** Keep tools sharp and clean for best and safest performance. Follow instructions for lubricating and changing accessories.

13. **Disconnect tools.** When not in use; before servicing; when changing blades, bits, cutters, etc.

14. **Remove adjusting keys and wrenches.** Form habit of checking to see that these are removed from tool before turning it on.

15. **Avoid accidental starting.** Don't carry plugged-in tool with finger on switch, and be sure switch is off when plugging in.

All the projects in this book can be built with a set of tools costing about fifty dollars. This figure is really a minimum that allows safe and responsible use of your tools. In addition to the basic shop, you might purchase additional tools for advanced work and you will be able to add tools to your shop without spending any money by building them yourself.

The basic tools of the apartment carpenter:

electric drill and bits
saber saw
screwdriver
claw hammer
clamps
squares
tape measure
Surform
mat knife
pencil
level

ELECTRIC DRILL

This is the most important tool in the apartment carpenter's workshop. While its primary purpose is drilling holes, with inexpensive accessories it can become a grinder for sharpening knives, a saber saw, a circular saw, a table saw and lathe, and even an air compressor for spray painting.

Electric drills are classified according to the size of their chuck—the radial vise which clamps the shaft of the drill bit. The best for apartment carpentry is a quarter-inch drill, which is usually smaller, lighter, faster, and cheaper than other sizes. If you need to drill larger holes, there are drill bits with shafts turned down to fit a quarter-inch chuck. Try to find a drill rated at 2.5 amps or greater. The speed of the chuck, measured in revolutions per minute, should be 1500 or 2000. Variable speeds and a reversing switch are not strictly necessary, though they are useful for drilling hard woods or masonry.

SABER SAW

Where the electric drill is flexible, the saber saw is specific: it cuts. Depending on the blade, it can cut wood, metal, cardboard, plastic, paper, or leather. The stroke of the blade is the measure of the saber saw. Some machines have a half-inch stroke (the distance the blade moves up and down), others a three-quarter-inch stroke. Short strokes at a high speed are best for wood cutting. The shoe—a wide flat plate connected to the housing of the saw—can be fixed or adjustable. In apartment carpentry, an adjustable shoe is often useful for making beveled edges on tabletops.

SCREWDRIVER

The screwdriver is your chief assembly tool. Look for a screwdriver with a large handle that's comfortable to hold. When you purchase a good screwdriver, also purchase a cheap one for opening paint cans, prying apart boards, chiseling away at wood surrounding a nail before pulling it out with a hammer, and all the other jobs which a screwdriver is so handy for, but which would wreck a good one.

The screws you will be driving will almost invariably be flat head screws, which can be driven flush or recessed into the wood when countersunk.

CLAW HAMMER

The major use of the hammer in apartment carpentry is in assembly, though screws and wooden pegs are much stronger and neater for finished work than nails. The hammer to use has a sixteen-ounce head and a wooden handle. As with any other tool, the most important criterion is feel—a good balance and easy energy flow between you and the work.

CLAMPS

Clamps grip and hold materials while you work. They come in many varieties—including C, bar, web, spring,

and corner—and are among the most useful devices of the apartment shop.

C clamps are C-shaped pieces of metal with an adjustable bolt threaded through one end of the C. They are classified by the maximum opening of the C, measured in inches; the size most useful to the apartment carpenter is four inches, which is large enough to grip two two-by-fours. You should purchase at least three of these clamps. The clamps you buy should look more massive than you think you need. They should weigh several pounds, with an adjustable bolt at least one-half inch in diameter.

Although other clamps are not necessary to the basic shop, you may find them useful as you continue building. Web clamps are useful for gripping round or oval objects. You may need a web clamp for building barrel-like projects, like planters. Some band clamps come with metal corners that protect the band from sharp edges, allowing the clamp to be used for picture frames.

Bar, or pipe, clamps are used when a wide board is being clamped and are adjustable from one to six feet. These are especially useful when gluing up a wide board from many narrow pieces. Corner clamps are used for gluing the corners of picture frames. Spring clamps can be very handy as quick grips. One type of spring clamp has movable jaws with teeth which dig into the wood, permitting the spring clamping of 90- and even 120-degree joints.

SQUARES

Both the carpenter's square and the adjustable square are useful in the apartment shop.

Carpenter's squares are L-shaped pieces of metal; the long arm is 24 inches long and 2 inches wide, and the short arm is 16 inches long and 1½ inches wide. Both arms are ruled in inches and eighths of an inch. The carpenter's square is used for layout and checking right angles.

The adjustable square has a body which holds a blade and a thumbscrew for tightening or loosening the blade. Some adjustable squares also come equipped with a level and a scribe. The blade is a removable steel ruler 12 inches long. Opposite the square portion of the body is a 45-degree angle for drawing miters.

TAPE MEASURE

Distances greater than 24 inches can be measured with either a folding rule or a tape measure. Folding rules have twelve 6-inch segments connected near their ends so they can unfold to 6 feet or fold back to 6 inches. They are nearly indestructible, but they are limited to 6 feet. A tape measure is a slicker, more modern device. The tape is wound, sometimes by spring, inside a metal case. When it is extended, it can be held at any length, either by lock or by friction. For apartment carpentry a 12- or 20-foot tape is fine.

SURFORM

A recent tool, the Surform, has characteristics of a rasp, a plane, and a gouge. It consists of a body with a handle on top and a pierced piece of metal on the bottom—it looks like a cheese grater and does to wood what a cheese grater does to cheese. Unlike a plane, it can work across the grain. It can round corners, plane surfaces smooth, and sculpt wood.

MAT KNIFE

A mat knife is a very handy apartment tool. Basically a thick razor blade with a heavy handle, it can be used to shave wood, cut veneer and cardboard templates, and do many other jobs requiring a strong, sharp knife. Some mat knives have retractable blades, others have blades that can be snapped off when dull, revealing a new sharp edge.

LEVEL

A level is a tool of exquisite precision. A level consists of a curved vial filled with fluid and a bubble held in a bar of wood, metal, or plastic that has at least one straight edge. The vial is marked with two lines, and when the bubble is between the lines the straight edge is level. Some levels have a second vial at right angles to the first to check verticals. Errors as small as .1 degree are immediately apparent with a level. Determining that your bookshelves are level is like hitting a bull's eye. What you may find in your apartment, however, is that the shelves are level but nothing else (the floor, the ceiling, the molding) is.

NAILS

The nails you will be using most often are of two types: finishing and common. Finishing nails have small heads and are used where you don't want nails to show. Because of their small heads, they can be driven, with the aid of a nail set, below the surface of the wood. Filling the tiny hole with plastic wood makes the nail disappear. The recessed nail also allows finishing operations, such as sanding to be performed without fear of snagging and tearing the sandpaper. And it is especially important when finishing is being done with a plane or other cutting tool.

Common nails have large heads and are generally heavier than finishing nails; they are used when appearance is not important. Nails are classified by type and size, the latter term frequently expressed in pennies, abbreviated, British style, by the letter "d". The term of "penny" to designate nail size came from the practice of selling nails by the hundred—so a hundred ten-penny nails would cost ten cents, two hundred, twenty cents, and so on. Nails aren't that cheap any more, but the term continues. To convert from pennies to inches, see the table in the Appendix.

SCREWS

While there are many kinds of screws, the apartment carpenter need use only one type: the flathead wood screw. This screw has three parts: the head, flat on top to lie flush with the wood, conical on the bottom to spread the force; the body, smooth so it slips through the outer board; and the screw itself, with deep threads to engage the wood fibers.

There are three things to consider when drilling for flathead wood screws. First, the hole initially drilled to locate the screw (called the pilot hole) should be small enough to engage the threads but not so small that the screw binds. Second, the body hole should be large enough for the body of the screw to drop into the hole, but not so large that the head drops through also. Note that the head hole is drilled with a bit called a countersink, and should be deep enough to allow the screw to lie flush. If you're wondering how deep to drill, try too deep rather than too shallow. This keeps the screws at or below the surface of the wood, avoiding snags. Several tool companies make special drill bits which drill all three holes in one operation. These bits are designed for particularly sized screws, and will say so on the package. In the Appendix is a table of pilot hole and body hole sizes for the different-sized screws.

Flathead wood screws are designated by two numbers: the body size (usually 4 to 14), and the total length (from ½" to 6"). When buying screws ask for, say, six two-inch number tens, or a box of number eights an inch-and-a-half long. While buying by the box of one hundred is cheaper in the long run, most hardware stores have rotating racks of blister-packed screws, useful in a pinch when you need just four more.

GLUE

Two glues are useful in apartment carpentry. The most familiar to everyone is white glue. Cheap, moderate working time, and fairly strong are its chief virtues. Less well known is a beige sort of glue sold under the trade names of Titebond and Carpenter's Professional Wood Glue. This glue is a delight. It sets up in a half hour, curing to full strength overnight. If you're laminating table legs this means gluing all four legs in three hours instead of four days. It dries without staining the wood, letting you clean the glue seam for finishing. This is important when staining, since nothing shows up through stain like a patch of glue. While it is more expensive than white glue, the little you use for each project is worth the extra expense.

OTHER TOOLS

You may have occasion to use nuts and bolts instead of screws. You will then require wrenches to hold the nuts and tighten the bolts. It's a good idea to stick to a few sizes of nuts and bolts. That way, you can buy appropriately sized wrenches instead of an adjustable wrench which tends to round the corners of nuts.

After you have assembled a project, you'll want to finish it. This means sanding the wood smooth so the grain shows clearly and then applying a finish of some kind. Finishes are liquids—paint, varnish, shellac, or oil—which are applied to the wood with a brush and then dry to seal the wood and produce a smooth visual texture. The tools of finishing are sandpaper, brushes, steel wool, and the finish itself.

Other tools which are handy to have around the apartment shop include an awl for marking the point when drilling, masking tape, and a compartmentalized case for holding small parts.

The kind of tools you buy will have a direct bearing on the quality of your work. While it is not necessary to buy the best, a certain minimum level of acceptance must be observed. The consequence of using shoddy tools is frustration, often resulting in abandonment of the project. To avoid this, keep to well-known brands of tools, or else follow the advice of a carpenter you know and trust.

MATERIAL

WOOD

To most people, wood comes from a lumberyard. Lumberyards do carry the widest selection of clean, new lumber, plywood, and other building materials. They also sell worked wood, such as moldings, and wood-like products, including particle board and Masonite. But, despite the variety of materials, most do-it-yourself carpenters tend to stick with pine boards. The most important reason for this is that pine, and only pine, is available in most lumberyards in a variety of widths and one standard thickness—¾-inch thickness and widths ranging from ¾ to 16 inches.

Lumber—and pine is lumber—is sold in simple-sounding dimensions: one-by-six, one-by-eight, two-by-four, four-by-four, and so on. However, you must remember that the "one" in a one-by-six, for example, stands for. one-inch thickness of lumber straight from the mill. After this rough-cut board is processed (by smoothing the surfaces and edges with planers), the "one" becomes three-quarters and the "six" is five and a half inches. Thus, two-by-fours are really one-and-a-half-by-three-and-a-halfs. As if all this weren't enough, small widths, such as one-by-threes, are notorious for being off dimension, for varying in width from piece to piece. A table in the Appendix gives the actual dimensions of lumber for the common sizes.

Pine wood has been used in this country for centuries. The early settlers prized it for its looks, workability, and availability. And for these same reasons it is widely used by apartment carpenters today. Since pine has been used for so many years—Early American furniture is almost exclusively pine—thousands of unknown craftsmen have left us a legacy of pine furniture designs; these designs have been published in many books.

While pine has been used for hundreds of years, plywood measures its lifetime in decades. Invented during World War II for PT-boat hulls, plywood has found its way into nearly every American home. A composite, plywood is composed of thin layers of wood, called plies (if very thin, veneer), sandwiched together with glue under heat and pressure to form large, flat, smooth surfaces. The grain direction of the plies alternate to equalize strength. Though any size is possible, plywood has been standardized into four-by-eight-foot sheets, the size universally available.

Plywood can be used anywhere surface is needed—table tops, bookshelves, doors, floors, and walls. It is not to be used where high strength in narrow section is required—table legs and loft supports, for example.

Unlike pine, in plywood the size you ask for is the size you get. It is available in thicknesses from one-eighth to three-quarters inches, in one-eighth-inch steps. Three-quarters is the size used most often in apartment carpentry, for bookshelves and desks, table tops and loftbeds.

Since lumber is growing scarce, ways are being found to utilize more of the log that has been cut. Chief among today's new products is particle board, also known as chip board or flakeboard, which is produced from wood chips bonded with glue under pressure. It is used for table tops, as a base for vinyl or asphalt flooring, and as a base for plastic laminates, such as Formica. Strong, smooth, extremely resistant to warping or changes in dimension, particle board is rapidly replacing plywood in inexpensive furniture construction. Its primary drawback is that it rapidly dulls cutting tools and cannot accept nails.

Another wood product of great utility to apartment carpenters is hardboard, known under the trade name Masonite. Made of wood fibers bonded under heat and pressure, hardboard is strong, dense, and durable. Perforated, it is sold as pegboard, and it is useful where strong, cheap surfaces are needed, such as the backs of bookcases or workbench tops.

Some lumberyards sell cabinet woods, especially walnut, mahogany, and maple, which are highly prized for their beauty, and are expensive when purchased as solid lumber. A less costly way to use these woods is as veneer, which is rarely available in lumberyards but is sold by specialty houses through mail order.

At the lumberyard, carry a tape measure when buying wood cut to size, so you can make sure all the pieces are correct before you pay for them. Also, you should know what to look for when inspecting lumber. The quality of lumberyard wood has fallen markedly in recent years, and consumers should be wary. Fortunately, the defects in wood are easily recognized. They are knots, warp, wind, and checks.

Knots

Knots are those dark, sappy, hard, circular defects in wood. Depending on how the log was cut, they can be round, oval, or extend completely across the plank. Formed when a tree branches, they are more frequent on younger trees, and their predominance in most lumber is an indication of overcutting by the lumber companies.

If the knots are tight, the board is usable where moderate strength is needed. Loose knots can be glued into the board, but its strength is reduced. Knots that extend across the board limit the use of the board to light loading conditions, such as part of a glued-up table top. Try to avoid leaving knots at the ends of boards. For bookshelves, table tops, panels, or drawers, the knots should not be more than two inches wide.

One method for increasing the strength of knotty wood is lamination, in which process several boards are glued together, not unlike plywood, to form a thick plank stronger than an equivalent thickness of solid wood. If the knots are at or near the edge of the wood, cutting them out may help. The basic test for solidity in wood is what I call the Whack Test: take the end of the board in both hands and whack it on the floor; it it doesn't snap at the knot, you can use it. (You probably won't be allowed to perform this test at the lumberyard, so avoid boards with wide knots.)

Warp

While knots are not usually detrimental to utilization of a board, warp—a bend in the wood caused by unequal drying—can destroy a project. To detect it, sight along a board with one eye. A dip, or a rise, or a curve off true indicates warp. Don't buy warped wood.

Wind

Wind (pronounced as in what you do to a watch) is warp in two dimensions. It ends up as a twist and is easy to recognize. Lay the board on a flat surface and press down on each corner in turn, then turn the board over and repeat the process. Any rocking proves wind, and unless you're carving a propeller, avoid it.

Checking

Checking, a separation of the grain which looks like a split, usually occurs only in thicker boards. It can sometimes be corrected by gluing and clamping, but severe checking down the length of a board diminishes its strength, making it nearly useless.

STREETWOOD

Lumberyards, however, are not the only possible source of wood for apartment carpenters. City streets bear new treasures every day. Furniture and appliances, often in need of only the most minor repairs, lie in heaps. And there is wood of every description, condition, and variety.

Hence, streetwood: any wood thrown away. A table top, slats from a box spring, wooden Venetian blinds, used two-by-fours near a construction site, broomsticks—in short, any wood on the street that's been put there as trash.

Wood is such a wonder material, almost every piece of streetwood could be reused. If you're not collecting materials for a specific project, seek the largest and longest pieces that you can transport and store. Then perform the Whack Test against the corner of a building. Besides testing for soundness, this helps you determine the board's workability. If the wood shows a dent, it's soft. To find out how soft, run your thumbnail (beware of splinters) along the grain—the deeper the groove, the more workable the wood. If the wood shows no dents, and your thumbnail makes no impression, it is a hardwood that dulls tools.

The soundness of the wood helps you determine its strength: as a shelf, will it support a hundred pounds of books or twenty cans of vegetable soup? Check a knotty board with a double Whack Test. (Knots are harder than the surrounding wood, so don't do the thumbnail test there.)

Since streetwood has been out in the weather, be doubly sure it contains no warp or wind. Refuse from a building under renovation after a fire will likely have both water and fire damage. The Whack Test will check for soundness in charred pieces. Water damage appears as wavy lines of dried soot. Plywood may loosen or separate with water damage. Solid cabinet woods (oak, maple, birch, walnut, cherry) are subject to splitting or checking after exposure to weather. Don't use any wood not in good condition. You can always find more. One thing this country's not runnin' out of is trash.

The surface is what you see first in a piece of wood. If you don't like its looks, check to see whether improvement is possible. Paint and varnish can be removed when not too thick. A scratched surface is difficult to repair, but doesn't impair strength.

Where to Find Streetwood

Since streetwood is any wood that has been thrown away, look around garbage cans near apartment buildings before trash-collection time. Here you can find furniture, the refuse of renovation, and the fine piles of junk new tenants remove from old apartments. If you like to make small boxes, the sides of bureau drawers are perfect: often oak or poplar, they've been protected within the bureau and are clear and smooth. Both woods take a fine finish.

While apartment houses are superb sources of well-surfaced materials, construction sites with contractors' boxes (those big green metal bins in front of buildings under construction or renovation) are rich sources of plywood and lumber, which includes the structural members such as two-by-fours and two-by-sixes torn from walls and floors during four-wall renovation.

If you are looking for small pieces of unused cabinet woods, like walnut or cherry, try visiting a furniture factory. One piano company in New York gives some of its "scraps" away to local high-school workshops. Smaller companies are often open to individual requests.

Crates are a good source of lumber, but only a few heavy goods are crated any more. Motorcycles from Japan come crated in a white mahogany, which when stained shows the full richness of its more colorful cousins. Glass companies' crates are an apartment carpenter's dream: two-by-fours eight feet long, one-by-tens, four feet long, two-by-sixes, and more. One glass company in Brooklyn has a sign on its gate: FREE WOOD. Any company that uses bar steel has crates it gets rid of.

Lumberyards themselves often have scraps they give away, likewise cabinet shops. When people move, they throw out a lot. If any of your friends are moving, offer to help them in exchange for their castoffs. (Bookcases often change hands at this time.)

While you can find streetwood any time, call the city's sanitation department to learn when local trash pickups are made, and precede the government's removal. You may also feel less criminal doing your scrounging during daylight hours.

Before you go stalking the streets, you should have justified in your mind exactly why using streetwood is a great idea. The first reason is economic: streetwood is free for the taking. On another level, you're not giving your money to Georgia-Pacific or any other lumber companies which pollute the air and water. You're also recycling existing resources, lessening the trash load on the environment. Finally, streetwood offers a greater variety of woods than you'll find in any lumberyard anywhere.

It isn't necessary to know everything about a trade to practice some of its skills. Most drivers can check the air pressure of tires. A smaller number can tune a car's engine. Still fewer can replace an engine. Finally, only highly skilled individuals can rebuild an engine.

In carpentry, hammering nails is the equivalent of checking the air. The skills and procedures in this book are all simple, never more difficult than changing a tire. And like changing a tire, carpentry requires the beforehand knowledge of the procedure. Having done it a few times helps too.

MEASUREMENT AND LAYOUT

Pencils

The pencil is a basic tool. It can be used to mark measurements, write them down, draw lines, figure bills of materials, or design new projects. Any pencil will do for figuring, but a No. 2 makes the most distinct line on wood. (Carpenter's pencils are too tough to sharpen.)

Ruler

A ruler has two jobs in carpentry: measuring lengths and drawing straight lines. The most useful is 6 to 18 inches long (15 to 45 centimeters) and can be made of wood, metal, or plastic.

When drawing a straight line through two or more points, it's important to remember that a pencil point has thickness and that, after a little use, it becomes rounded. In this condition, it cannot fit into the corner between the work and the ruler. To allow for this when drawing the line, place your pencil at the first point and bring your ruler up to it. Using the tip of the pencil as an axis, rotate the ruler until the edge of the ruler is the same distance from the second point as it is from the first. Check to see if this allows for the thickness of the pencil point by placing the pencil at the second point and drawing a short line. This line should go through the middle of the point on the board—if it doesn't, move the ruler until it's correct.

Distances along a straight line are lengths, and are always measured between two points on the line. To measure a length, line up the ruler on the first point and swing the ruler around that point until it just touches the second point. Don't use the end of the ruler to measure from—it can be inaccurate.

Carpenter's Square

The primary use of a carpenter's square is checking squareness, a very simple procedure. To check an inside angle, such as the meeting of two walls of a room, put the corner of the square into the joint. Rest one edge of the square against one surface and see if light shows between the edge of the square and the other surface. If the gap is at the end of the square, the angle is greater than 90 degrees; if the gap is at the corner, the angle is less than 90 degrees; if there is no gap, the angle is exactly 90 degrees, and thus square. The procedure is similar for checking exterior angles, but the interpretation is reversed. Place the inside edge of the square on one surface of the angle. Slide the square into the angle until the other inside edge touches. Check for gaps. If the gap is at the corner, the angle is greater than 90 degrees; if the gap is at the end, the angle is less than 90 degrees.

To familiarize yourself with the square, try testing various parts of your apartment—doors, doorframes, floor-to-wall joints, table legs, etc.—for squareness.

Since it is large and heavy and won't slip on the paper, the square is also useful as a layout tool for drawing straight lines and making angles. Angles other than right angles are often needed, and they too can be drawn with the square. This method of constructing angles is a very tidy procedure and is far more accurate on large work than a protractor. Establish your base line and mark the vertex, or corner, of the angle to be drawn. Check the table for the angle you want to draw, and position the long arm of the square on the line, with your vertex point at the value shown in the table. On the short arm, mark the value shown in the table with a short line and a point. Draw a line from the vertex to the point, and you have an angle drawn with an accuracy of better than one part in a thousand. (For the 72- and 75-degree angles reverse the arms.)

The table of angles in the Appendix includes all commonly used angles and all the angles used in this book.

Compass

For apartment carpentry, a dime store compass is sufficient. If the arms are tight enough to maintain constant radius, it is the perfect tool for drawing circles and marking dividers.

Circles will be used in rounding the corners of tables and desks, or making wheels for toys. The point of the compass marks the center of the circle, acting as an awl if the center is to be drilled. When drawing a circle, lean the compass ahead of the pencil point, dragging the pencil around the circle, so the point won't grab on rough grain.

For dividers, a compass can mark off equal lengths along a straight line or around a circle. You can thus mark off a series of equally spaced points for adjustable legs on a chaise longue or an adjustable clamp.

Templates

A template is a pattern fashioned of thin wood or cardboard used to mark shapes on wood. Templates are useful for repetitive work, like cutting notches in several boards. Also, a greater degree of care can be expended on a template, insuring the same level of accuracy all through the job.

One method of making a template is to draw the shape on paper, glue the paper to thin wood, then carefully cut the shape. Thin wood is easily shaped with sandpaper, allowing hairline precision.

CUTTING THE WOOD
Saber Saw

As you prepare work for cutting, remember these three things: select the right tool for the job (in this case, the blade), measure twice—cut once, and clamp your work.

The thickness of the wood and the types of cuts and angles affect your choice of a blade. For example a two-by-four needs a blade with seven teeth to the inch, with set to the teeth. Plywood splinters easily, so a 32-teeth-to-the-inch metal cutting blade should be used. For very smooth cuts, you can use a hollow-ground blade. In some cases a bevel cut requires a longer blade than a square cut. There are wide blades available for straight cuts and narrow blades for curved cuts; the narrow blades have a smaller turning radius in the cut. For large-radius (greater than 3 inches) outside-curve cuts, you can use a wide blade. Sharp angles in a cut, such as right angles, demand backing and filling (a process described more fully later) and require a strong blade. The right choice of a blade can make the difference between a fast cut and a slow burn.

Measure first when laying out the work. Measure again just before cutting. Make sure you're not cutting a piece you don't need, or cutting it too short (a *very* common mistake), or cutting into the work side rather than the waste.

Clamping before cutting paces your work, giving you time to think. You can shorten a board with a saw, but you can't lengthen it. Once you develop a system for clamping, your work will be faster, cleaner, safer, and more accurate.

Sawing to a Line

Since a pencil line has dimension, all of the pencil line should be on the piece saved, so your saw cut will be entirely in waste wood. A few Xs on the waste side of the line drawn when laying out the piece will reduce your confusion before cutting.

There are three steps to a saber-saw cut: entry, the cut itself, and exit. Entry is the process of starting the cut at the right place on the board. To enter a cut, hold the saw in one hand, finger off the trigger. Place the front of the shoe onto the edge of the board, with the saw tilting slightly forward. The blade should not be touching the wood at this point. Place the thumb of your other hand on the shoe, forward and to one side of the blade—not touching the blade or the blade clamp, nor projecting in front of the blade. Spread and press the fingers and palm of this hand firmly on the board, well away from the blade and the line.

With this grip you have two-hand control of the saw. Start the saw, press down on the front of the shoe with both hand and thumb, and touch the blade to the wood on the waste side of the line. Move the saw from side to side until you have the blade on line. Now rotate the saw, bringing the shoe parallel with the board, and move the saw forward. You can keep your thumb on the shoe while cutting; this gives greater stability and less vibration to the operation.

For the cut itself, press down with hand and thumb and guide the *blade* along the line. Don't worry if the edge of the shoe or the body of the saw isn't parallel with the cut. Often the saw blade doesn't sit exactly parallel with the shoe, but this doesn't impair its cutting ability. Keep your head forward and above the cut to guide the blade and to blow sawdust away from the line. Remember to bear down firmly on the shoe, which puts all the force of the motor into cutting, not vibration. The tougher the cut, the firmer your grip and the more pressure you apply. If the cutting is really tough going (short of smoking), you can press the saw forward with your thumb. This pulls the blade into the wood. If you were to push forward with your saw-holding hand, the tendency would be to twist the blade, and possibly snap it.

If your cut starts to smoke heavily, turn off the saw—and wait for it to stop before setting it down. Unplug the saw and set aside, allowing the blade and motor to cool. Change to a sharper blade or one with fewer teeth.

When the cut is seven-eighths done, you must think about exiting. If you simply continued cutting, the unsupported board would snap off before being cut, leaving a rough or jagged edge. One method for exiting the cut is to remove your thumb from the shoe and grip the loose piece with that hand. Another method has an assistant hold the board while the cut is completed. A third method is to clamp the board at four points, two on either side of the cut.

Ripping

Ripping is narrowing a board by cutting it down its length, with the grain. The saw is guided along the board a fixed distance from one edge parallel to it. This can be done by eye, using a line marked on the wood, or for greater accuracy, with a fence—a board (½-¾ inches thick, 2-6 inches wide) with a known straight edge. The fence is clamped or nailed to the piece to be ripped and the saw shoe is guided along the straight edge. (This method works only if the blade of your saw is parallel with the edge of the shoe.) Remember that the fence is *not* at the line of cut but some distance (the distance from the blade to the edge of the shoe) from it. On most saws the blade-to-left-edge and blade-to-right-edge distances are different.

Crosscutting

Crosscutting, the process of cutting across the grain, can also be done by eye or with fence. Table saws and hand saws have separate blades for ripping and crosscutting, but saber saws have combination blades to do both.

Chamfering

Chamfering means beveling edges—that is, cutting them at an angle. When a board is used as a table leg, sliding it across the floor could cause it to split. Chamfering the board where it meets the floor prevents this. Chamfering can be done with the Surform plane or sandpaper backed by a block.

Pocket Cuts

Pocket cuts are cuts started and finished without reaching the edge of a board. They are made by standing the saber saw on the front of its shoe with the blade nearly horizontal, and with the motor running, rotating the blade into the wood. When the blade breaks through, normal cutting can then continue. It often helps to nail a block in front of the saw before the cut so the saw has something to bear against while cutting. If the cut is made too quickly the wood will grab the blade, which may snap it, so work slowly.

DRILLING

Drilling a Hole at a Point

Select the drill bit and insert it in the chuck of the drill. Tighten the chuck with the chuck key. The point to be drilled must be indented with an awl or nail before drilling, because the cutting tip of a drill bit is not a point, and there must be room for the bit to seat before cutting. (Failure to indent the hole may cause the bit to travel and start cutting in the wrong place.) Hold the drill at right angles to the board with the bit at the point to be drilled and drill. Always drill slowly. If the board is thick, back the bit out of the hole every three-quarters of an inch or so to clear the chips. To prevent splintering, clamp a backing block for the bit to drill into. When using spade bits (flat bits used for drilling large holes) always use a backing block. When this isn't possible, drill until the tip of the bit just pokes through and finish the hole by drilling from the other side.

Drilling for Flathead Wood Screws

Flathead wood screws require three holes—two cylindrical and one conical:

1. The pilot hole (so named because it pilots the screw through the wood) marks the position for the other two holes and allows room for the screw to cut its threads without splitting the wood. It is drilled first, if possible, through both boards to be joined simultaneously. If you can't drill both boards together, drill through one board and use it to mark the holes on the other.

2. The body hole (or shank hole) allows the screw to slip through the first board and should be large enough so the screw does fall through the board up to its head.

3. The countersink is a conical hole that allows the cone-shaped head of the screw to seat itself flush with the surface of the board. When the screw is tightened, the head of the screw is pulled against the countersink, forcing the two boards together. The countersink is drilled with a special bit called a countersink. The angles of flathead woodscrews and countersinks are standardized so one countersink should fit all screws.

Some devices on the market use a specially shaped bit which can drill all three holes in one operation. These are useful when you can clamp the boards that will be screwed so the holes can be drilled in one operation; otherwise these gadgets have little utility.

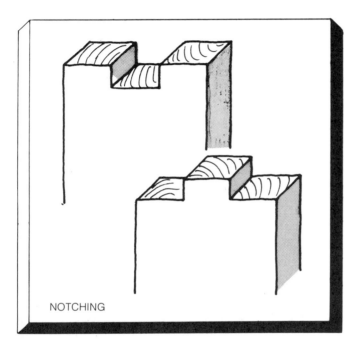

NOTCHING

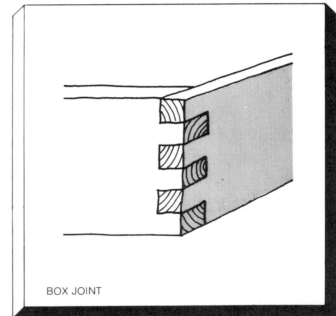

BOX JOINT

NOTCHING

Cutting a notch, or notching, is the basis of most saber-saw joints. A notch is a squared-off space cut from a piece of wood; its counterpart, a square tooth at the end of a board, should neatly fit the notch. Notches can be narrow and deep or they can be wide and shallow; a board can end in one, two, or many notches, as in the box joint.

If you have many joints to cut you can make a jig for marking the depth. Nail a scrap piece of wood to another scrap piece an inch wider. Both should be at least as long as your boards are wide. Use 1½-inch nails (four-penny).

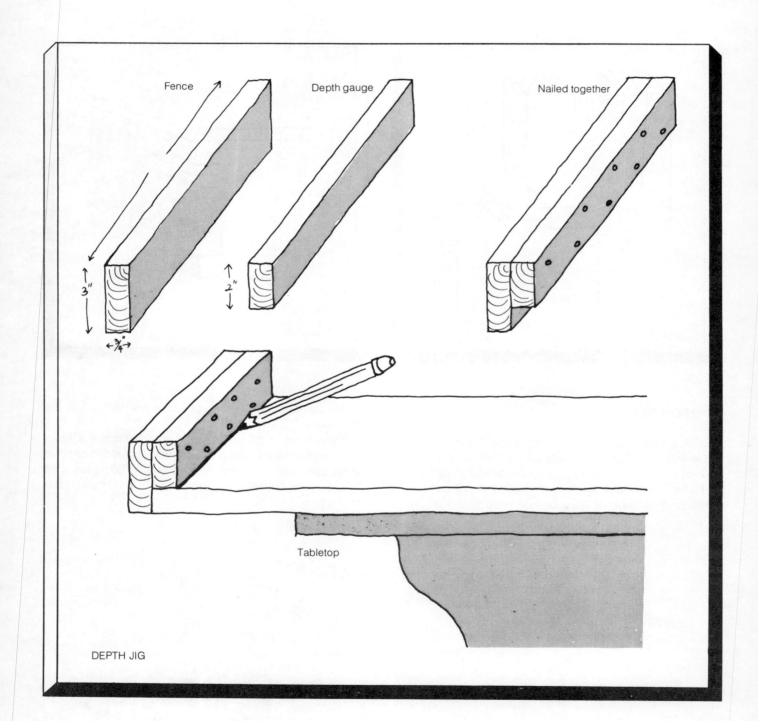

Fence

Depth gauge

Nailed together

3"

¾"

2"

Tabletop

DEPTH JIG

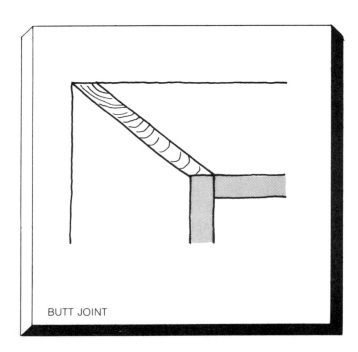

BUTT JOINT

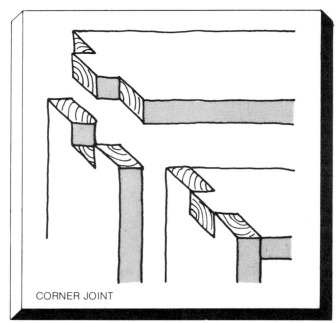

CORNER JOINT

Corner Joint

The corner joint illustrated here is superior to a simple butt joint because it can be nailed together from two directions. The corner joint has two parts: the notch and the tooth. Start by clamping the board you're working on to the table. Mark both pieces for depth, using a scrap piece of wood the same thickness as the boards. This insures the proper depth of the notch and shoulders—what's cut away to make the tooth. (Pine and hardwoods often have irregular thicknesses, like 25/32 inches, so measuring can be difficult and inaccurate.)

Having decided how wide your tooth will be (one-third the width of the board is good; other dimensions are useful in special circumstances) subtract this distance from the board width. Set *half* this distance on the adjustable square and mark the shoulder cuts for the tooth on each side of the board. This puts the tooth in the center of the board. Cut out the tooth and use *it* as a template for the notch. Here orientation is important.

Keep outside edges out and inside edges in. A strip of masking tape on or near the edge with the words "top shelf outside edge" or "top out" should make construction less confusing.

Next draw the notch on your board, using the tooth you've cut as a template. (This insures that the lines are drawn on the piece saved.) Enter the cut, cutting to the corner of the notch (Figure 2). Back up about three or four blade widths, saw still running, and shave off some wood, again cutting to the line. You want to cut out a triangular space in the waste wood that will have room for the blade at the top. This may take three or four passes (Figures 3–6), backing up and cutting forward.

When you have shaved out enough room for the blade turn off the saw, waiting for it to stop, and re-insert the blade for the cut across the notch (Figure 7). Cut to the corner, but now turn off the saw. Start cutting again on the third line. Cut to the corner, where the

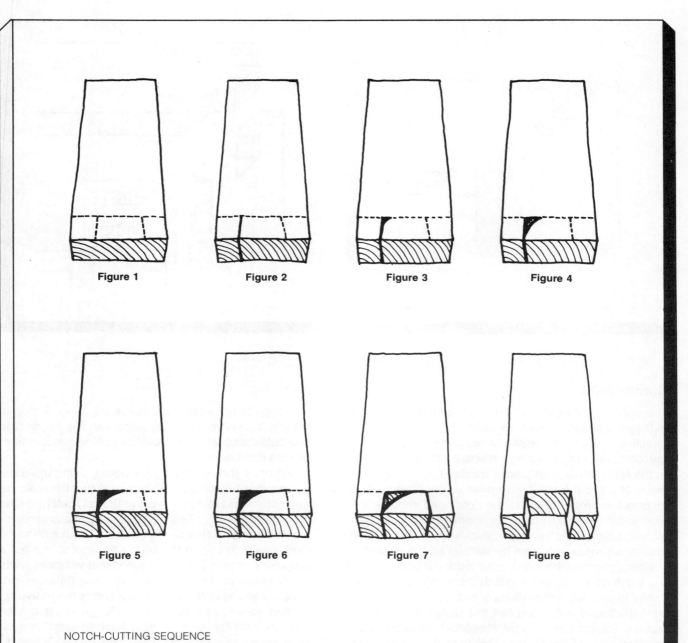

Figure 1   Figure 2   Figure 3   Figure 4

Figure 5   Figure 6   Figure 7   Figure 8

NOTCH-CUTTING SEQUENCE

waste piece should drop out (Figure 8). Check the corners of the notch to be sure they're smooth and square. You can use the saw to shave away rough areas. Making a cut directly into the corner a little ways sometimes helps the tooth-notch fit.

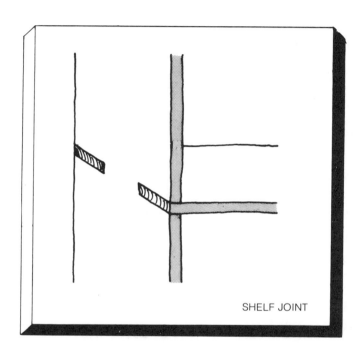

SHELF JOINT

Shelf Joint

This joint gives the shelf full support front and back, making it strong enough for records, blocks, or books. It's an elegant joint, very satisfying to make and to see in use.

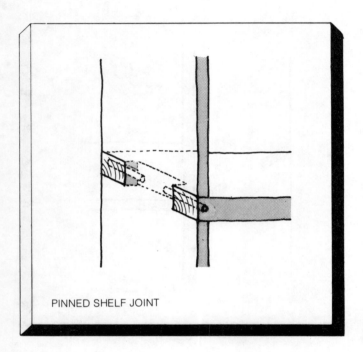

PINNED SHELF JOINT

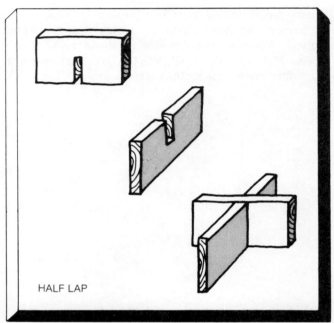

HALF LAP

Pinned Shelf Joint

Here the teeth of the shelf are narrower and are pinned (nailed, screwed) to the upright with dowels. Like the corner joint, it has the strength that fastening in two directions gives.

Half-Lap Joint

The half-lap is probably the most useful joint in apartment carpentry. Two boards half-lapped and held together with a top make a table. A half-lap in a corner—a corner lap—has the advantages of two-direction nailing, making it nearly as strong as a corner joint, and being easier to cut. Change the dimensions of the corner lap and you have the end lap, one of the strongest joints in framing. An excellent joint for two-by-fours, the half-lap

CORNER LAP

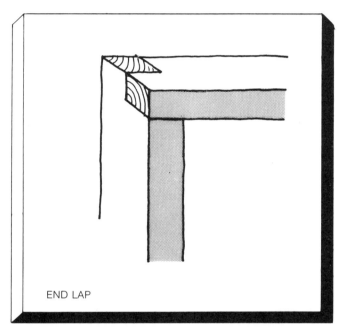

END LAP

has many applications in bed frames and lofts. Finally, half-laps, unlike most joints, can be adjusted before assembly if cut incorrectly. If they are too large (loose), thin slips of wood (called shims) can be glued in; if too small (tight), more wood can be removed.

The half-lap has two parts—two identical parts, both notches. Because there are two parts, orientation is critical. First decide which pieces will have notches opening "out" and which "in." Shelves traditionally have smooth fronts, so they'll open in; uprights then must open out. Next comes position, which is also critical, and it's best to mark for both sides of the notch on the board.

The width of the notch is the thickness of the board fitting into it. Use a scrap of wood of the same thickness to mark the width of the notch. Better yet, make a simple jig which you can use to mark all three sides of the notch at once. Cut the notches, and cut to fit. A trifle loose is better than too tight. Ideally, the joint should be snug, requiring a gentle tapping for assembly.

2
PROJECTS

BENCH ONE

Around a dining-room table, against a wall, or under a loft, benches can seat one or many. They thus serve a useful function in an apartment: conserving space that would otherwise be taken up by several chairs. This bench is a good first project. All pieces are cut at the lumberyard and the assembly uses glue and nails.

Begin by nailing the braces to the legs. The corners of the two pieces being nailed offer natural right angles with which to line up the two pieces. Nail one brace into the two legs, then flip this assembly and nail in another brace. Use glue in the joints for a tighter bench. Make both pairs of legs in this way. Mark the rails for the legs with the adjustable square beginning 9 inches from each end. Prenail the rails, position them on the glue-ready legs, and drive the nails home. The two braces and leg joint provide enough gluing surface to insure a strong, sturdy frame. After you have nailed on the seat, the bench is complete, ready for finishing or for use raw.

Bill of Materials:

4 – 2″ × 4″ × 15¾″	legs
4 – 2″ × 4″ × 11½″	braces
2 – 2″ × 4″ × 36″	rails
1 – 1″ × 12″ × 36″	top (seat)

about 30 – 8d common nails (2½″)

about 20 – 6d finishing nails (2″)

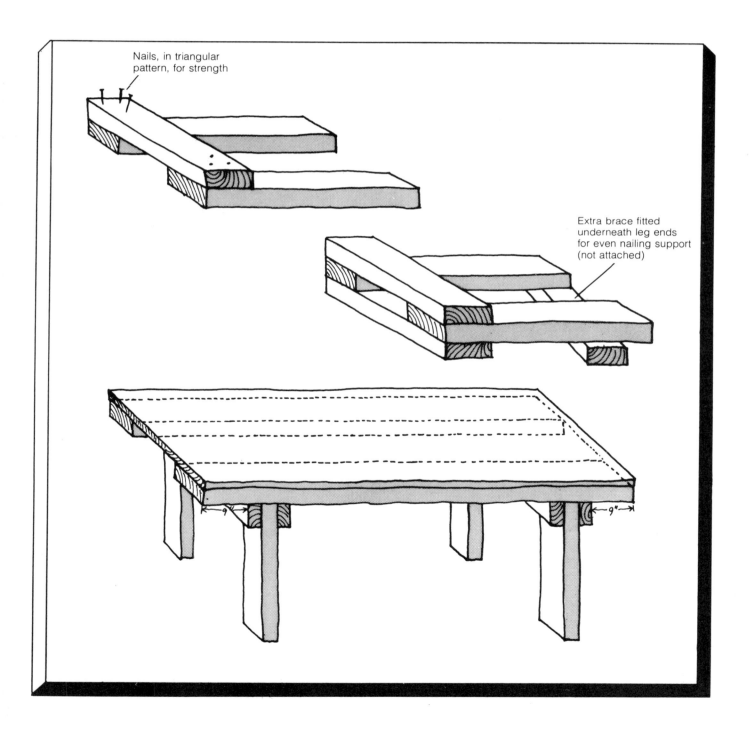

Nails, in triangular pattern, for strength

Extra brace fitted underneath leg ends for even nailing support (not attached)

9" 9"

FOUR-BOARD BENCH

The outline for the legs is developed using a paper template. Cut a sheet of paper to the size of a leg board (newspaper does nicely) and fold it in half. Draw the outline for one foot on the sheet. Cut it out, unfold, and transfer the foot detail to one of the leg boards. Clamp the two legs together and cut the feet. Chamfer the feet so they won't splinter. Draw two parallel lines ⅜ inch on either side of the center line of the leg boards on both sides of both boards. This is where the rail will be attached.

Prenail the legs and attach the rail. Be sure the rail is nailed inside the lines. When both legs are attached to the rail, stand the assembly on its feet and nail on the seat. Be sure to center the seat on the leg assembly.

With the bench completed, you might try cutting the corners of the seat round, using a drinking glass or jar lid for the template. Sand all edges smooth, with both Surform and sandpaper. Finish with a penetrating stain that allows the wood to breathe, and thus age gracefully.

Bill of Materials:

2 − 1″ × 12″ × 17¼″	legs
1 − 1″ × 12″ × 36″	seat
1 − 1″ × 8″ × 28½″	rail
6d finishing nails (2″)	

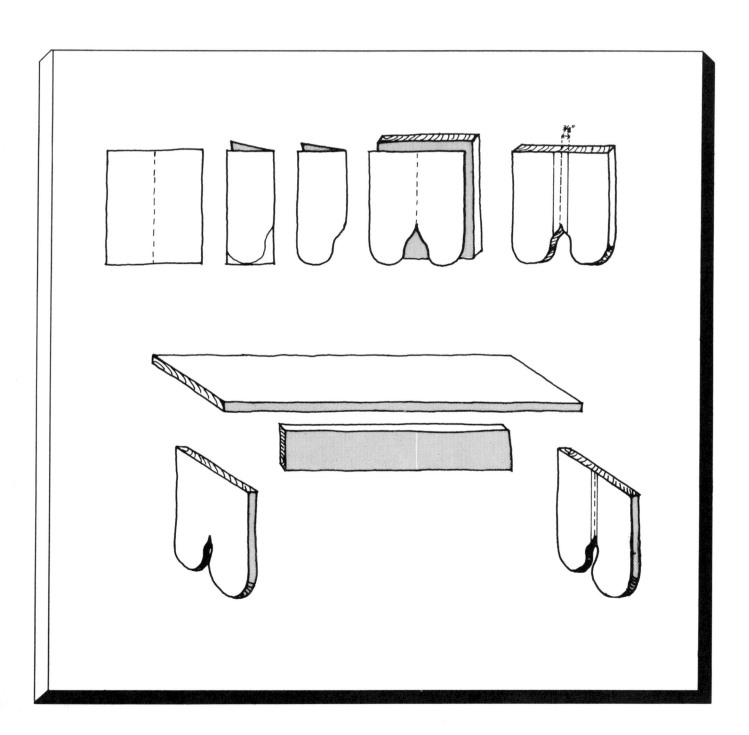

FIVE-BOARD BENCH

Commonly used as a footstool design in colonial New England, this bench is simple to construct.

Begin by notching the leg boards to accept the rails. Draw and cut the leg detail, clamping the two boards together for symmetrical cuts. Construction is straightforward. Nail and glue the rails to the legs, nail and glue the seat to the leg rail assembly. Knock off the corners of the seat with the Surform or saber saw, plane with Surform and sand smooth. This bench should last for fifty years.

Bill of Materials:

2 — 1″ × 12″ × 17¼″	legs
2 — 1″ × 8″ × 30″	rails
1 — 1″ × 12″ × 36″	top (seat)
6d finishing nails (2″)	
glue	

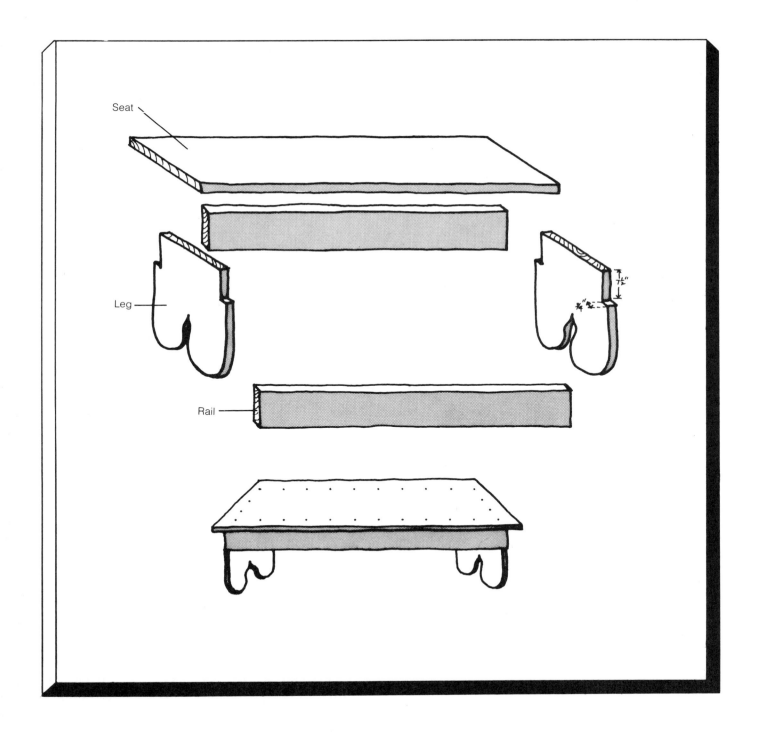

X-BRACE TABLE

This table is based on a popular colonial design. It is strong, light, and easy to build. The legs are built up (laminated) from two pieces of ¾-inch pine, which allows you to construct a wide half-lap joint without chisels.

Begin with a jig for the angled ends of the leg pieces. To give it a 45-degree angle, cut a right triangle with 5-inch sides from ⅛-inch Masonite. Use a scrap 1 inch by 2 inches for the fence of the jig and nail it along the short side of the triangle (see Figure 1). With the jig mark all the leg ends for cutting. Cut all the legs, then glue and nail the pieces as shown in Figure 2, using one of the one-by-four boards for accuracy in positioning. Glue and nail the braces for the legs. At your option, you can cut notches in the legs to accept the stretchers (the long boards which hold the legs apart), or simply nail them into place. With someone holding the table steady, nail the top to the legs 9 inches in from an edge, using finishing nails and ample glue. Countersink the nails for filling later with wood putty. When both legs are attached, flip the table on its top. Glue and nail the wide stretchers in place near the table top. Flip the table back on its feet and glue-nail the narrow stretchers into position. When the glue is dry, the structure is complete. Fill in the nail holes with wood putty and sand smooth. This is a good table for staining and varnishing, especially if the top is of high quality (veneer plywood or solid wood). One possibility for a table top is two one-by-eighteen boards 5 feet long edge-glued for a top 60 inches by 35 inches.

Bill of Materials:

8 — 20″ × one-by-fours	legs
4 — 43½″ × one-by-fours	legs
2 — 42″ × one-by-sixes	stretcher
2 — 42″ × one-by-threes	stretcher
1 — ¾″ × 36″ × 60″	top
4d finishing nails (1½″)	
glue	

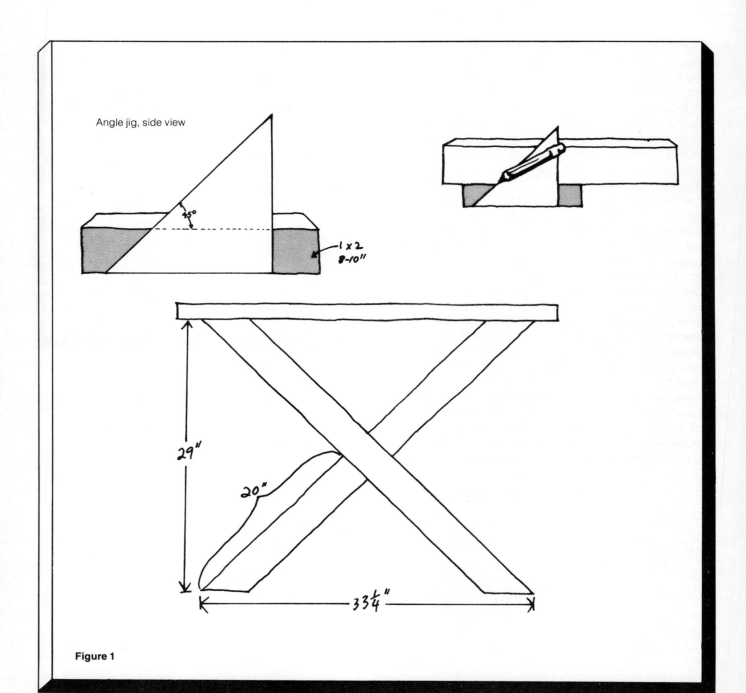

Angle jig, side view

45°

1 x 2
8-10"

29"

20"

33¼"

Figure 1

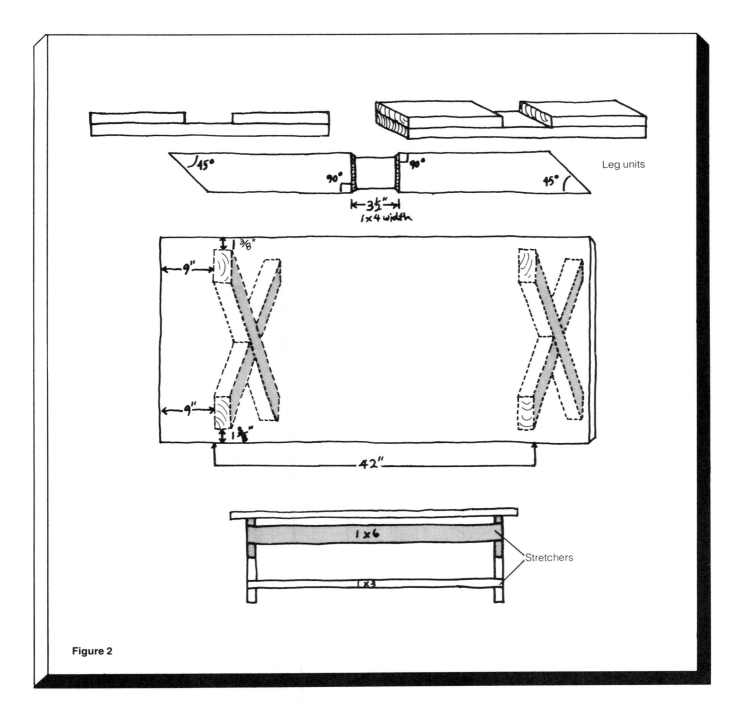

45°

90°   90°

Leg units

45°

1⅜"

|←— 3½" —→|
1x4 width

9"

9"

1⅜"

42"

1x6

Stretchers

x3

Figure 2

JACK TABLE

Mark the center point of the two-by-two legs. Make the line all the way around the piece. Using one piece as a gauge, mark off two squares on the pieces, as shown in Figure 1. Spread glue on these squares.

Clamp the three struts together, square to square, with 4-inch C clamps, as shown in Figure 3. Use your adjustable square to make sure all the pieces are at right angles to each other. When the glue is dry, remove the clamps and set the piece on the ground so three ends touch the floor with the other three ends parallel. With a Surform and a good eye, take down the top corners until you have a thumbprint-sized flat on each end. Flip the jack and do the same for the other three ends.

Place the top on the floor, put the jack on it, and mark the position of the legs where they touch it (Figure 4). Make sure the leg ends are equidistant from the edge of the top. Nail a 1-inch finishing nail halfway into these three marked points and pull them out. Taking new nails, tap them into the holes—head side in, point side out. Spread generous amounts of glue onto the flats of the leg ends and tap the legs into the nails until the glue oozes from the joint. Carefully flip the table standing and weight it with thirty to fifty pounds of books or bricks.

Bill of Materials:

3 — 36″	two-by-twos (See Note)
1 — 24″	round top
glue	
2d finishing nails (1″)	

Note: The pieces of two-by-two *must* be square in cross section for the table to work. Do not let the lumberyard sell you two-by-fours cut in half unless they cut them square. This will mean two cuts for each two-by-four. Measure the pieces with a ruler and demand square pieces.

When the glue is dry, invert and sand the ends touching the floor round. Finish as desired.

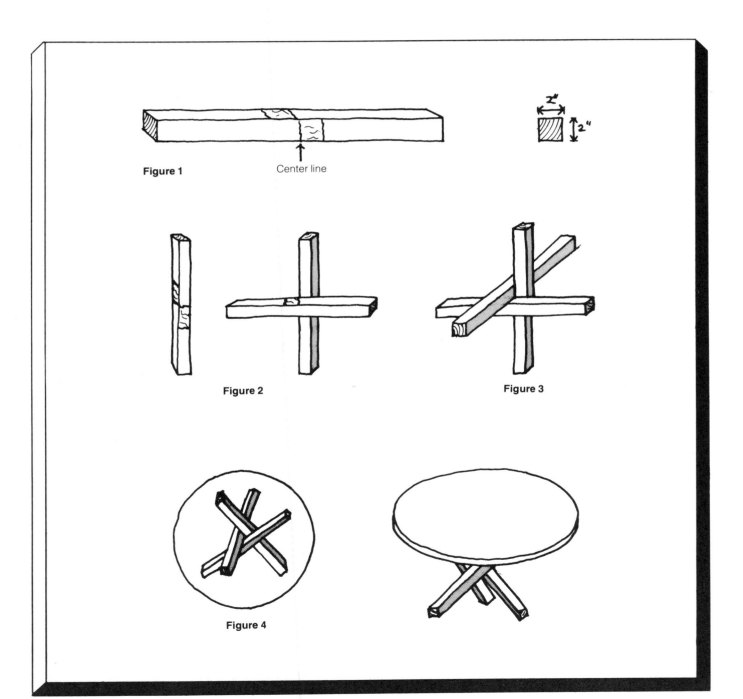

Figure 1

Center line

2"

2"

Figure 2

Figure 3

Figure 4

CHINESE TABLE

The most notable feature of this design, adapted from a table exhibited at a museum, is the use of glued-up legs that form a bridle for the table rails without skillful cutting. The technique has applications in desks, chairs, and any other furniture with legs.

The table has three groups of components: legs, rails, and top. The rails and legs are of pine, for light weight and beauty; the top can be of plywood, butcher block, or a glued panel of pine.

Legs

The legs are assembled with glue and clamps. Spread glue evenly on the inside surfaces of the joint (see Figure 1). Apply enough glue so it just oozes from the seam when the clamps are tight. Use scrap pieces of ¼-inch plywood as pads between the wood and the clamp faces so you don't mar the surface. Clamp the pieces long-short-long (see Figure 1) with your three 4-inch clamps. If you're using Titebond glue (which sets in an hour and cures overnight), you can unclamp the leg after an hour and move to the next. When all the legs are glued, they can be cleaned of glue drips and smoothed with the Surform plane.

Rails

Mark off the rails 8 inches from each end (see Figure 2). Use a leg as a gauge to mark its width on the rail (see Figure 3). Draw the center line on the rails (Figure 4). Using the lid of a two-pound coffee can, or its equivalent, mark out the curves on the rail (Figure 5 and 6). Cut the rails with the saber saw, smoothing the edge with the Surform. The short rails can be marked and cut the same way (Figure 7).

Assembly

First glue the legs to the long rails with the 4-inch clamps, making sure the legs are square with the rail. Next glue the short rails to the legs, as shown in Figure 8. It may help to nail a one-by-two across the legs temporarily to keep them standing during this part of the assembly. When the glue is dry you can remove the one-by-two.

Bill of Materials:

2 — 1″ × 6″ × 48″	long rails
8 — 1″ × 3″ × 27¾″	legs (outside pieces)
4 — 1″ × 3″ × 22¼″	legs (inside pieces)
2 — 1″ × 6″ × 24″	short rails
1 — ¾″ × 36″ × 60″	top
10 corner braces 1″	
20 screws to fit brackets ⅝″	
glue	

Spread some newspapers on the floor and put the good side of the top down on them. Mark four points on the underside of the top 6 inches in from each end and 5¼ inches from the long edges (Figure 9). These are the four outside corners of the long rails. With an assistant, lay the leg-rail assembly onto the underside of the top. Place the corner braces along both rails, two on the short rails and three on the long rails, ten in all. Their placement is not critical but they should be evenly spread.

Slip a piece of cardboard under a corner brace and mark the screw hole on the rail with an awl or nail. Screw the brace to the rail (screws this size don't need drilled pilot holes, just deepen the awl hole). Do this for all ten braces. This elevates the braces slightly, causing them to pull against the screws, and drawing the top and rail tightly together. Tap the awl through the screw hole in the corner brace and screw the braces to the top. With the Surform, bevel the edges of the legs where they touch the floor to prevent the ends from splintering when dragged around. Stand the table on its legs, and it's ready for finishing.

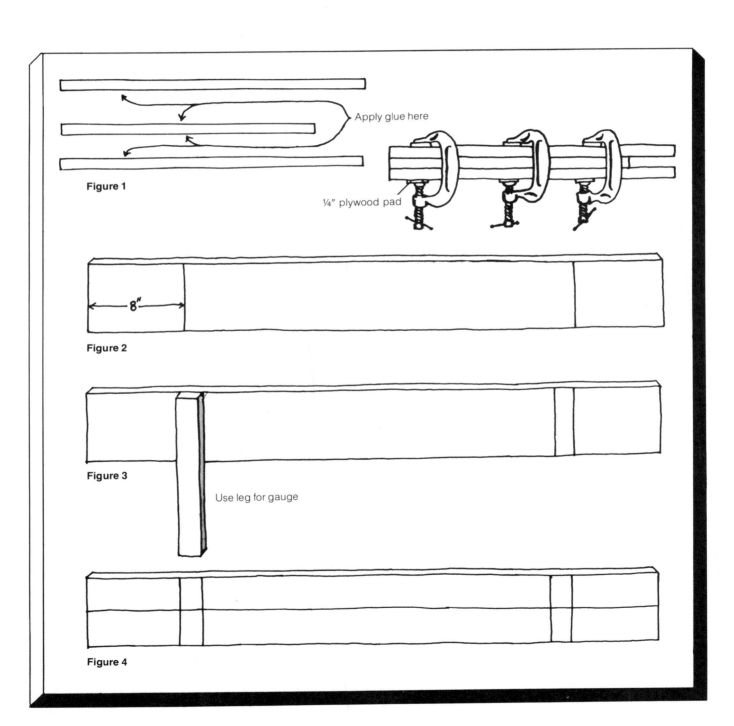

Figure 1

Apply glue here

¼″ plywood pad

Figure 2

8″

Figure 3

Use leg for gauge

Figure 4

42

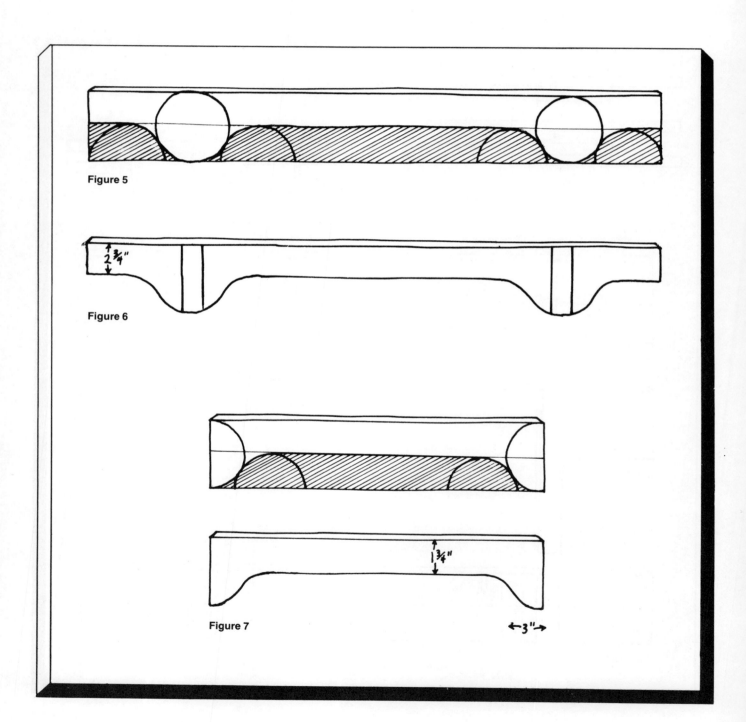

Figure 5

2 ¾"

Figure 6

1 ¾"

Figure 7

←3"→

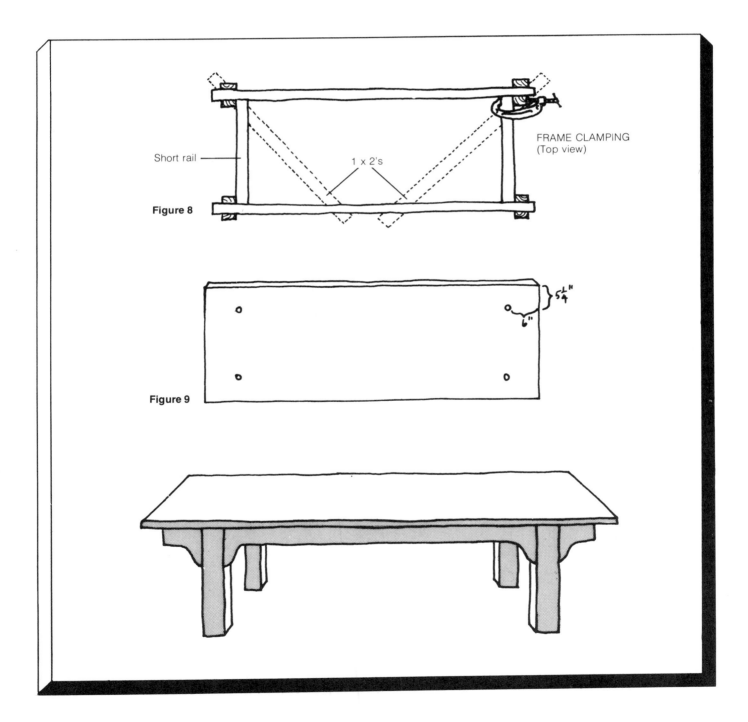

Short rail

1 x 2's

FRAME CLAMPING
(Top view)

Figure 8

$5\frac{1}{4}''$

$6''$

Figure 9

CHEVRON DESK

This may be the simplest table of all.

Glue-nail the brace to one of the legs along the short side ¾ inch in from the edge. Glue-nail the two legs together, nailing the braced leg into the other leg, and the other leg into the brace (two-direction nailing). With finishing nails, glue-nail the top to the L assembly. Cut the corners of the top round for a smoother-looking surface. This is a good one to paint. If you use plywood, paint the edges with a diluted solution of glue and allow to dry before painting, to hide the edges.

Bill of Materials:

2 — ¾″ × 28″ × 32″   legs

1 — ¾″ × 36″ × 48″   top

1 — 2″ × 2″ × 28″   brace

4d finishing nails (1½″)

glue

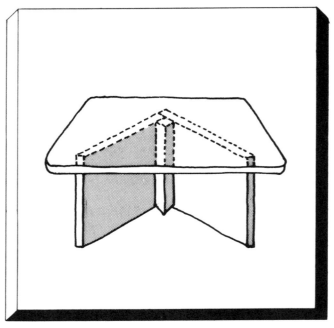

TRESTLE DESK

This desk can be assembled, sanded, painted, and ready for use in an afternoon. A perfect height for desk work, it doubles as a dining-room table for one or two. When built with a 36-inch top (available already cut at some lumberyards), it can seat four for dinner. It also makes a dandy table to work on during future projects.

Start by gathering the materials. Although I built this desk entirely from streetwood (walnut plywood from old headboards), you can purchase all the materials cut to size at a lumberyard.

The only saber-saw work in building the desk is in cutting the legs. Use a jar lid to draw the arcs for the feet as shown in Figure 2. Cut one leg and use it as a template for the other leg. While each leg is clamped, use your Surform or a piece of coarse sandpaper wrapped around a scrap to sand the edges of the plywood where they will touch the floor. Take off a lot, enough so the outer plies won't touch the floor (see Figure 3). This insures that those plies won't catch and splinter when you slide the desk around. While you're sanding, round the other edges (except the top) to avoid catching splinters on your clothes later.

Assembly

On the two legs, mark a center-line, running from the top edge 7 inches in. Now draw two more lines, one on each side of the center line and 3/8 inch away. Make these lines 7 inches long also. Do this on both sides of the legs. Your center line should be 8½ inches from either long edge of the leg. On the inside this shows you where to glue, on the outside of the leg where to nail. Put the leg on the floor, outside up, and hammer five 2-inch nails about 1 inch apart and starting 1 inch from the end down the center line. Try not to let them poke through. Do this for both legs.

Apply glue to the end of the rail and to the inside of one leg between the lines. Stand the rail on end, glue side up, and position the leg on it so the rail is between the lines and flush with the top of the leg. It helps to have an assistant at this point. Drive one nail part way— check to see that all is flush and even; drive a second nail—check again; drive the rest home. The first nail acts as an axis allowing adjustment, the second nail fixes the piece in position.

Flip the rail-leg assembly to glue and nail the other

Bill of Materials:

2 — ¾" × 17" × 26¾"   legs (plywood)

1 — ¾" × 17" × 21¾"   top (plywood)

1 — ¾" × 7" × 20¼"   rail (plywood)

1 — ¾" × 30" round   top (particle board)

one pound common nails (2")

one pound 3d common nails (1¼")

glue

leg, using the same procedure. Stand the leg-rail assembly on its feet and set in a safe place. Now take your square top and draw a center line the long way. Draw two lines the short way, each 3/8 inch from each end. Prenail this piece along the lines using 2-inch common nails 2 inches apart, again starting 1 inch from the edges. Bring out the leg-rail assembly and apply a generous coat of glue to the top edge. Keep a wet rag around to wipe off the excess. Position the square top— it should just fit—and drive the nails home, using the check once, check twice method. Have your assistant sit on the top while you nail.

Now take your round top and, with the good side down, draw a diameter. Spread some glue on the top of the leg-rail top assembly and place it on the round top, lining up opposite corners of the top rectangle on the diameter line. Position it so the distance from the corner to the round edge is the same for both opposite corners. Nail the two tops together with the 1¼-inch common nails. Use as many as you like, but a minimum of eight on each side of the rail. Try a pattern from Figure 4 if you are unsure. When the glue is dry, the piece is ready for sanding and painting.

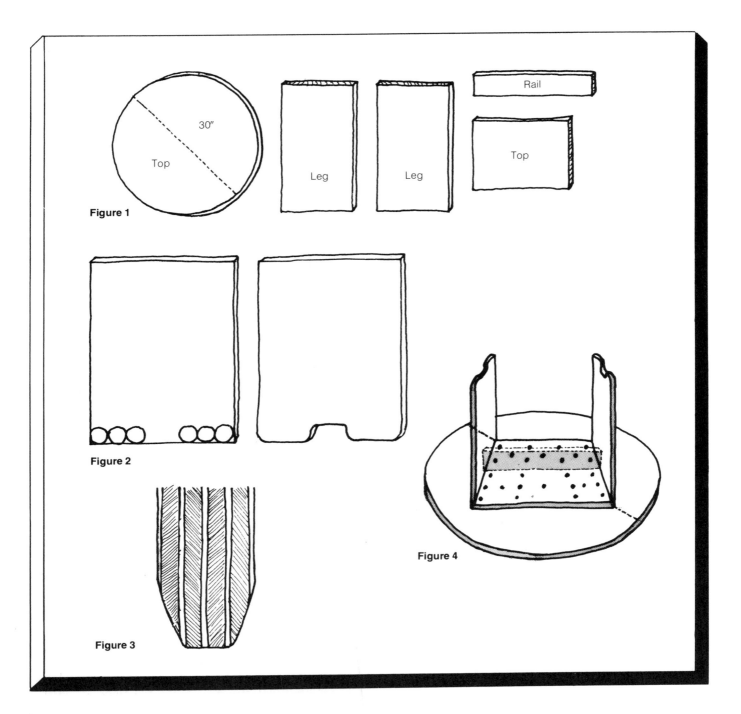

Figure 1

30"

Top

Leg

Leg

Rail

Top

Figure 2

Figure 3

Figure 4

SAW TABLE

Often saber-saw cuts, especially in thin or thick wood, require table support of both work and cutoff. This saw table is both simple to build and adaptable in use. On it you can cut off long boards, make notches in all sizes of lumber, and cut intricate scroll work. The table will fit in any closet and can serve as a workbench when clamped to an existing table.

The legs are built on a mass-production basis. Measure and mark the first piece of a set and use it as a template for the rest. With all pieces cut, make four L units by nailing the 7¼-inch pieces into the 6¼-inch pieces (Figure 1). Glue-nail two L units together, long side to long side (Figure 2). Glue-nail the long boards, capping the T (Figure 3). Position the two Ts (as shown in Figure 4) and glue-nail. Nail on the table boards and the structure is complete (Figure 5). Round the corners and Surform the edges to subdue the harshness of the particle board. Or leave the saw table unsanded and unfinished—it should withstand great abuse. To operate the table, clamp the baseboard to an existing table with a C clamp. Clamp the board to be cut to both tables, with the cutting line in between. Leave enough room for the saw between the clamps, especially when cutting notches and scrollwork.

Bill of Materials:

4 — 6¼″	one-by-threes
4 — 7¼″	one-by-threes
2 — 14″	one-by-threes
2 — ¾″ × 10″ × 16″	particle board
1 — ¾″ × 18″ × 28″	particle board
4d nails (1½″)	
glue	

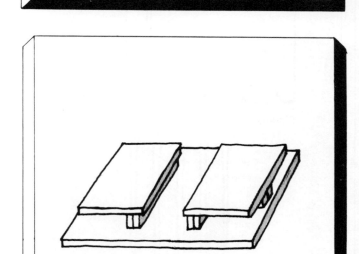

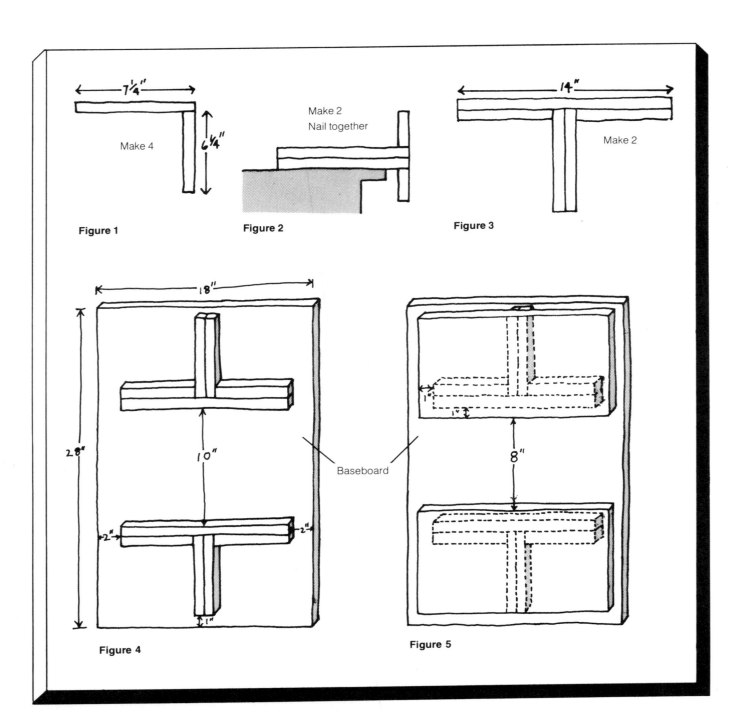

Figure 1

Make 4

7¼"

6¼"

Figure 2

Make 2
Nail together

Figure 3

14"

Make 2

Figure 4

18"

28"

10"

2"

2"

1"

Baseboard

Figure 5

1"

1"

8"

FALSE-DADO BOOKSHELVES

This four-shelf (one paperback, three hardcover and large paperback) bookcase requires no home sawing. Using just hammer, nails, and glue (you can even skip the glue), with wood cut at the lumberyard, you can build strong, freestanding bookcases in an afternoon.

Prenail all the spacers in this pattern or make up your own (Figure 1). The pattern should have left-to-right and top-to-bottom symmetry for even pressure on the glue, and the outer nails should be an inch in from the board's edge.

Place an upright good side down on your nailing surface (table, floor). Working from the bottom up, spread the bottom spacer with glue and nail it to the upright, one edge flush to the front edge of the upright. This will leave a ¼-inch gap at the back (see Figure 2). Place your gauge on the top edge of the bottom spacer (see Figure 3) and fit a glue-spread spacer up against it and flush with the front of the upright. Nail it down. Repeat with gauge and spacer all the way up. Stand this upright up where you can see it, spacers showing.

You will be following the same procedure for the other upright but here caution is indicated. Do *not* duplicate the first upright. Instead, make a mirror image of it. Start from the bottom, again good side down. When placing the spacers, if flush edge was toward you on the first piece, turn it away from you on this one. Remember to keep your spacers matched 7½ with 7½ and 10½ with 10½. If in doubt, make a direct comparison. When both uprights have been spaced out, you can install the shelves.

Fit two shelves into two slots on an upright (Figure 4). Stand this up on the ends of the shelves and nail the shelves in place, taking care to make the front edges flush. Do this for all the shelves, gluing and nailing in place, then flip over. After spreading glue in all slots, fit the shelves in. This can be tricky. If they're tight, hammer the upright by pounding on a scrap piece (the gauge) placed on top to avoid marring the wood. When all the way in and flush, nail home. Turn the shelves on their front, fit the back in place, and nail home.

Bill of Materials:

2 — ¾″ × 7¼″ × 7½″ spacers (one-by-eights ripped down ¼ inch)

6 — ¾″ × 7¼″ × 10½″ spacers

2 — ¾″ × 7¼″ × 2¼″ bottom spacer (see Note)

2 — ¾″ × 7½″ × 45″ uprights (one-by-eights)

1 — ¼″ × 36″ × 44″ back

1 — gauge cut from shelf wood 2-3″ wide

5 — ¾″ (?) × 7¼″ × 36″ shelves

4d finishing nails (1½″)

glue

Note: At the lumberyard measure the actual thickness of the shelf boards—if they are ¾ inch the bottom spacers are 2¼, if the shelves are thicker than ¾ inch subtract fives times the deviation from 2¼. (For example, if the boards are 13/16 inch, that minus ¾ is 1/16, five times 1/16 is 5/16, and 5/16 from 2¼ equals 1-15/16. The bottom spacers, in this example should be 1-15/16 inches tall.) Test the bottom spacer at the lumberyard by stacking the spacers on the upright end to end and check that the shelves fit into the space left. (See Figure 7.)

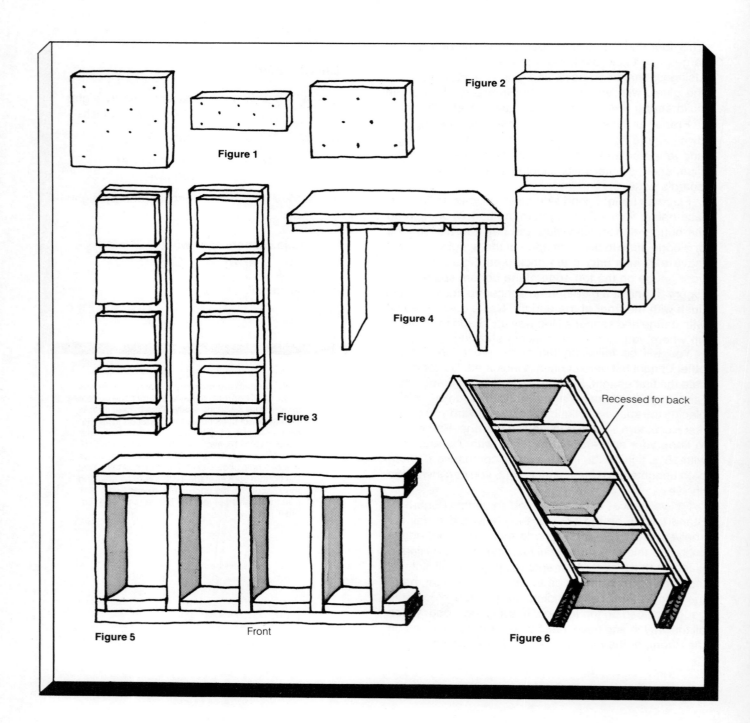

Figure 1

Figure 2

Figure 3

Figure 4

Figure 5

Front

Figure 6

Recessed for back

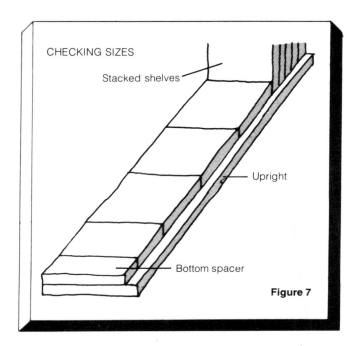

CHECKING SIZES

Stacked shelves

Upright

Bottom spacer

Figure 7

PAPERBACK BOOKCASE

If you have lots of paperbacks, this one's for you. It is light, strong, and easy to assemble, and the total cost of materials is under five dollars. Most of the cutting is done by the lumberyard, and this project can be completed and in place in an afternoon.

Construction begins with a template. To make it, draw the slot depth line 2¼ inches in from a long edge (Figure 1). With another piece mark the end notches (Figure 2). Measure in from each end 7-13/16" and mark the cross-line with the adjustable square. Again mark the notch width with another piece of ¼-inch plywood.

Cut the template carefully, using another piece to check the slots for fit. With the template cut, mark another piece with it and stack four pieces together with a clamp. You can cut through this stack with the saber saw, saving much work. Repeat this stack-cutting with the three other pieces and sand all slots to remove splinters.

The pieces go together the way cardboard spacers in liquor boxes fit together—slot to slot. Assemble four pieces with glue, as shown in Figure 3. Now glue and nail on the top piece, making sure the shelves that will hold the books have their slots in back with the uprights' slots forward (see Figure 4). In this way the uprights hold the front of the shelves and the back holds the back of the shelves, giving full support front to back.

Flip the assembly, glue and nail on the bottom board. Glue and nail on the sides. In the corners you will be able to nail in two directions, further strengthening the piece. Turn the case on its face and glue-nail the back. Be very careful when nailing into the shelves because you have such a narrow target. A good method for

Bill of Materials:

8 — 4½" × 24" × ¼"   plywood (sides and shelves)

1 — 24" × 24" × ¼"   plywood (back)

four ounces finishing nails (¾")

nailing is to pound the nails halfway in (maybe a half dozen at a time) and check inside the case to see if they've poked through and missed the shelves. Pull out any that have and drive the rest home. The squareness of the lumberyard cut will insure the squareness of the whole piece. Finish as desired. This is an especially good project to paint, which will cover up glue drips. The books in place will help to hide any wide slots, and plastic wood can fill any glaring gaps in front.

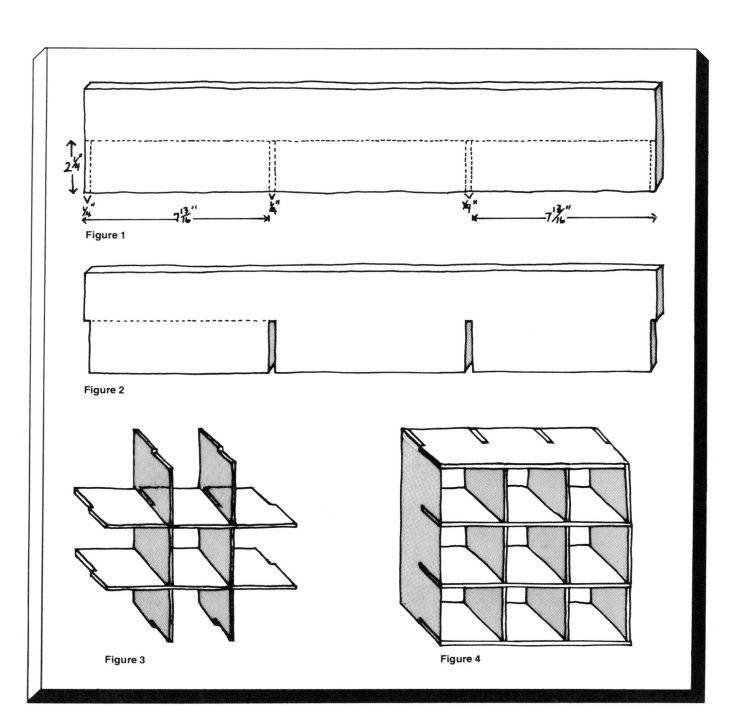

Figure 1

Figure 2

Figure 3

Figure 4

Mark one side piece as a template for the rest. Begin by drawing the line establishing the width of the notches with one of the other boards (Figure 1). Next, draw two lines, each 3¾ inches from the other two sides (Figure 2). This takes care of any local variation in wood width, for all the difference is taken up by the center notch. Mark with Xs the pieces to be removed (Figure 3), then carefully cut the notches. Your piece should look like Figure 4—sort of a sandwich with three bites taken out of it.

Place this template on the piece to be marked, with the grain running parallel. Draw the outline of the template (Figure 5). Again draw the lines establishing the width of the notches (Figure 6). Now X out and cut the *opposite* of the template drawing. If you outlined corner notches, cut center notch. If you outlined a center notch, cut corner notches. In this way, you can follow the edge of the pencil line without cutting into it. Otherwise you would destroy the line. (If you imagine drawing the lines with a paintbrush, this might be clearer.) What you're aiming for is close-fitting joints, and this method should assist you. Don't hesitate to check the joints against each other for fit.

The sides of the cube go together like a three-dimensional jigsaw puzzle, or a chain of acrobats standing on each other's shoulders, round and round. Apply glue generously in the joints and nail in two directions. A good method of assembly is:

1. Apply glue at one joint to all concealed surfaces, as in Figure 7.
2. Hold up the other end with another side.
3. Nail the glued joint together from the top side.
4. Glue and nail the other end of the topside piece.
5. Flip the assembly, apply glue to both joints.
6. Nail from topside.
7. Rotate and nail from other direction.
8. Repeat step 7 three more times.
9. Glue and nail on back, checking for squareness.

Bill of Materials:

4 – 1″ × 12″ × 11¾″ sides

1 – ¼″ × 11¾″ × 11¾″ back

about 30 - 6d finishing nails (2″)

about 15 common nails (1″)

This cube can also be assembled with screws, dowels, or clamps without nails, but nails are fastest. If you want no nails showing, use finishing nails and sink them below the surface with a nail set. Fill in the holes with wood filler and sand smooth when dry.

This size cube will hold hardcover books. Change the dimensions for paperback bookshelves, record holders, magazine storage, and so on. If you want to assemble several cubes into a wall unit, try building two or three different sizes with outside dimensions on a module, or common multiple. For example, one-to-two-to-three, as in Figure 8, or two-to-three, as in Figure 9.

This method of construction is not limited to cubes. The basic joint is adaptable to any box, as in Figure 10. The chief advantage to a cube is that all sides are identical, simplifying layout.

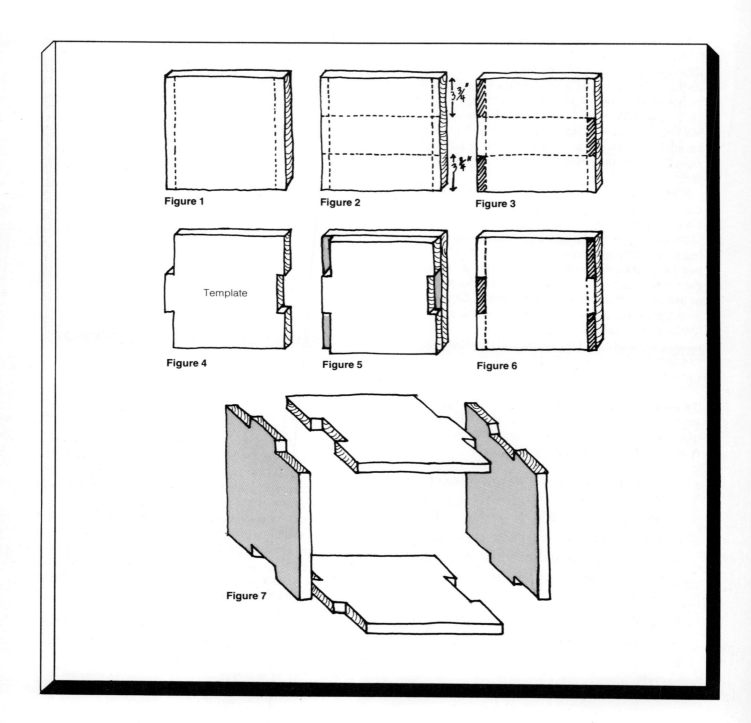

Figure 1

Figure 2

Figure 3

Template

Figure 4

Figure 5

Figure 6

Figure 7

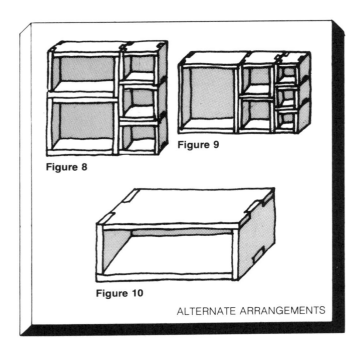

Figure 8

Figure 9

Figure 10

ALTERNATE ARRANGEMENTS

PORTABLE BUNK BED

Although it is portable, this bunk bed is sturdy enough to stand years of abuse by growing children. After the frames are built, a half-hour's screwdriver work is sufficient for assembly or disassembly.

With thirty-two notches to cut, a template is vital for repeated accuracy. Make the tongue of the template of ⅛-inch plywood or hardboard for durability, and the fence of a scrap 1 inch by 2 inches. Glue and nail the tongue to the fence with ¾-inch nails. Take care that all is square.

Mark and cut four two-by-threes to the dimensions shown in Figure 1. Test the first few notches for fit and, if necessary, adjust the template. The eight crossbeams are next to be measured, marked, and cut. The notches are one "two-by" width (1½-inches) from the ends of the beams (which also happens to be the template width).

Decide on the heights of the two beds and mark the uprights at these distances, less the ¾ inch of the bed floors. Lay two uprights on the floor and set a notched rail across them, the ends of the rail flush with the outer edge of the uprights, and lined up with the height marks (Figure 2). Be sure the notches are facing up toward the top of the bed. Glue and nail the rail to the uprights using Titebond glue and two 2½-inch common nails. After repeating this procedure for the top rail, flip the assembly and nail the uprights to the rails, again using two nails per joint. This should insure an extra tight connection. Repeat this entire procedure for the second frame. When finished, the frames should look like Figure 3.

Place a frame on the floor rail side down. Slip a scrap piece of two-by-three under the ends of the uprights (Figure 3). Glue and nail the railing piece to the frame with 2½-inch nails. This puts the railing on the outside of the uprights for greater room on the upper bunk. Do the same for the other frame.

The floors of the bunk bed are cut from plywood sheets. Purchase whole sheets and have the lumber-

Bill of Materials:

10 — 2″ × 3″ × 76″		frame uprights and rails
8 — 2″ × 3″ × 45″		crossbeams
2 — ¾″ × 45″ × 76″		floor
2 — ¾″ × 45″ × 20″		headboards (left-over from cutting floor)

glue

32 - 8d common nails (2½″)

16 screws – 1½″ #10

yard cut them to size. At home, cut a 1½-by-2½-inch rectangle out of each corner to fit around the uprights and provide a shelf on both sides of a standard twin mattress.

With the frames built, the cross beams and floors cut, assemble the bed. The cross beams notch into the rails of the frame, and the floors rest on the beams. Attach the headboards with 1½″ #10 screws, drilling pilot holes with the headboards clamped into position. Three screws on a side are sufficient. With all pilot holes drilled and the boards marked for position ("top bunk—top out," for example), remove the headboards and drill the body holes and countersinks. Attach the headboards to the bed with screws only (Figure 4). This should stiffen the bed substantially. With the addition of a ladder, the structure is complete. Surform and sand all exposed edges, especially plywood edges, and finish as desired.

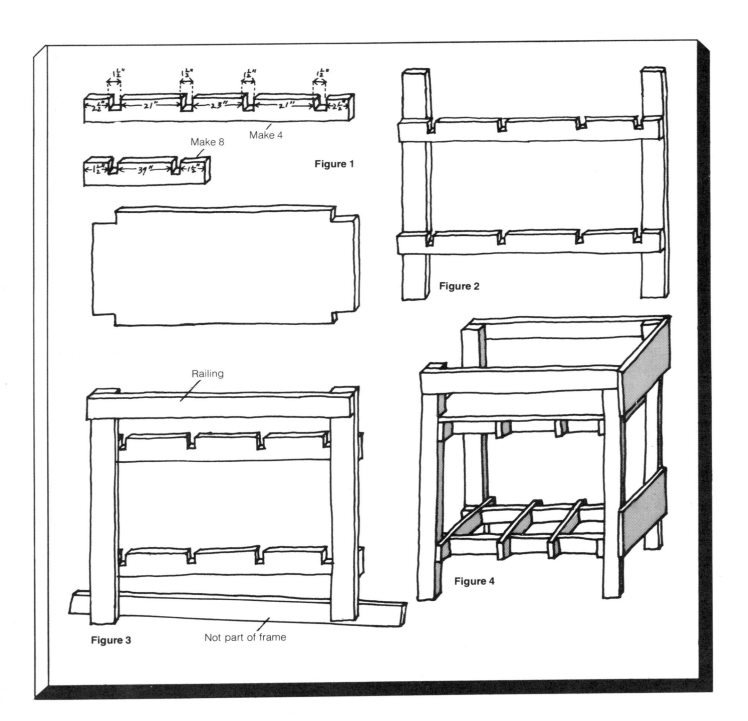

Make 8

Make 4

Figure 1

Figure 2

Railing

Figure 3

Not part of frame

Figure 4

LADDER

This ladder, suitable for a bunk bed, can also hang from a loft or lean against a wall. Simple to build and sturdy in use, it is another example of the "false dado" joint.

Begin construction by making a template for the risers, as shown in Figure 1. Mark all the risers, then cut them. Cut the bottom ends of the rails at the same angle (75 degrees). Determine whether the ladder will lean against the wall or hang from the bunk bed or loft. If hanging from the edge of the bunk bed or loft, place one of the rails into its eventual position, taking care to keep the cut end flat on the floor. Mark where the loft touches the rail. From this point draw a notch (see Figure 2) so the ladder can rest snugly up against the bunk bed or loft. This notch also enables the ladder to be attached to the bunk bed or loft with screws or dowels. If the ladder is to lean against a wall, cut the rail at a 15-degree angle (as in Figure 3) at the wall end.

After all the cutting is completed, you are ready for assembly. Starting from the floor end, glue and nail the risers to the rails. Leave a gap between two risers the thickness of a step. Use a step to set this gap. When one rail is completed, do the other, making sure it is a mirror image of the first. The rails, when placed next to each other, should look like Figure 4.

Glue two steps into their slots on one rail about 3 feet apart. Flip this over so the steps are standing on end. Tap one nail into the rail above the step. Center the step in the slot and drive the nail home. Drive in the other nail and repeat the procedure for the rest of the steps. When the steps are attached to the rail, flip the assembly over. Apply glue to the ends of all the steps and to the slots on the other rail.

Bill of Materials:

2 — 7′ two-by-threes   rails

8 — 18″ two-by-threes   steps

16 — 8-7/16″ one-by-three   risers

about 100 - 6d common nails (2″)

about 40 - 10d common nails (3″)

glue

Now comes the tricky part. Tap the other rail into place, with all the steps slipping into their respective slots. Once this is done, nail the rail to the steps, centering each one. Wipe away all the glue drips with a wet, almost dripping rag. Wipe away the water with a moist, nearly dry rag. This should prevent the glue from showing through the finish. Place the ladder into position and screw into place. This ladder can be used with any loft or bunkbed.

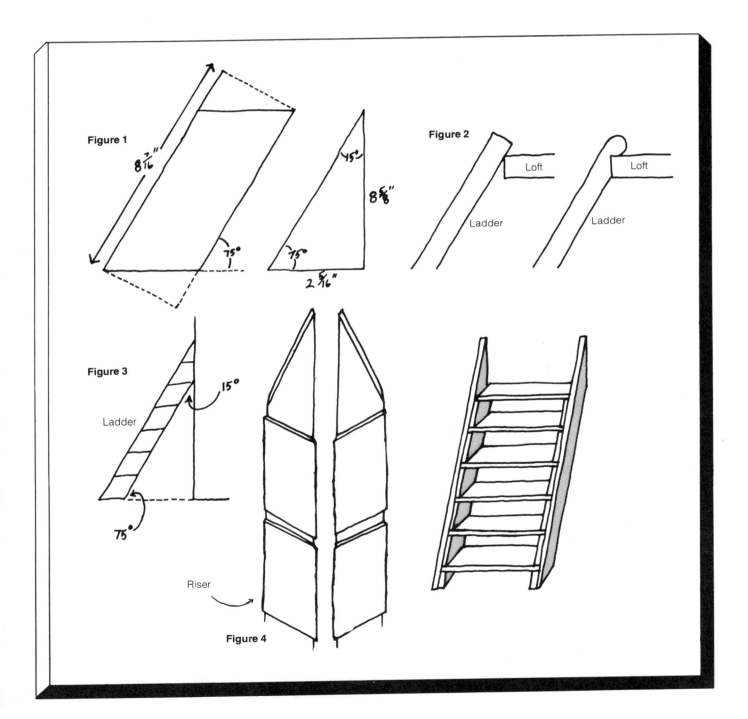

Figure 1

Figure 2

Figure 3

Figure 4

Walls are the overwhelming fixture of the apartment landscape. As an apartment carpenter, your expertise will be judged by how cleanly and securely you can attach things to walls. Following are descriptions of the various walls you are likely to encounter and the methods appropriate for solving particular problems.

PLASTER-AND-LATH

Plaster-and-lath walls are most commonly found in older apartments, and, if you use the correct methods, they are fairly easy to attach to. A typical interior plaster-and-lath wall has five layers: plaster, lath, stud wall, lath, and plaster. The stud wall consists of wood timbers, usually two-by-fours, standing upright, and roughly the same distance apart. The studs are covered on each side by a layer of lath. Lath is wooden strips, about ¼ inch by 1½ inches in cross section, nailed horizontally about ½ inch apart on the studs. The lath is covered by a scratch coat of plaster, which fills the gaps between the lath, and a finish coat of plaster, which is then painted.

There are two good ways to hang from plaster-and-lath walls—screws in studs or toggle bolts. Putting screws in studs means first finding the studs, which can be surprisingly elusive, hidden as they are. The only surefire method of locating studs involves drilling cores: using a small diameter bit, drill into the wall where you'd like a screw to hang. If you find the going slow and steady and bring up wood shavings, you've found the stud. Slow going followed by a hollow feeling means you've missed it; move over an inch or two and redrill. After you've located all the studs that will support the board, mark their locations on the board. Drill the board for screws, starting with pilot holes. Place the board in position on the wall and drill through the pilot holes into the wall. Remove the board and drill it for body holes and countersinks. This procedure should insure that the holes in the board match the holes in the wall.

Toggle bolts go between studs in the hollow wall. Be sure to drill a pilot hole to test for studs before drilling the larger hole for the wings of the toggle bolt (⅝ inch for a 3/16-inch toggle bolt). Otherwise you might come upon the stud, and drill a hole too large for any screw. About

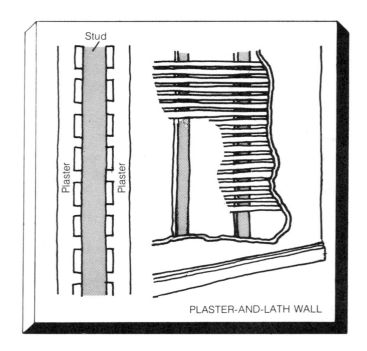

PLASTER-AND-LATH WALL

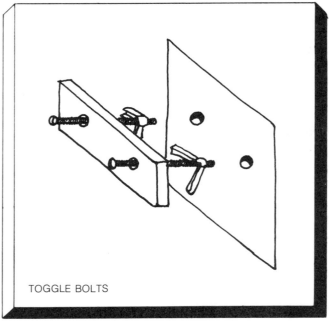

TOGGLE BOLTS

the only way to correct this is to drill a new hole. The toggle bolt must be holding something—a washer, a board—between its head and the wall, so the wall is tightly clamped and the hole is covered. The major use of toggle bolts is to secure boards to walls upon which shelves are supported.

SOLID PLASTER

Several methods have been devised for attaching to thick plaster: plastic anchors, fiber plugs, and lead plugs. All depend on friction and the procedure for all is the same. Locate the hole, drill for the plug or anchor, seat the plug, drive the screw in. The body of the screw forces the sides of the plug against the hole in the plaster, which snugs the fit. On very old or waterlogged plaster, this crumbles the sides of the hole, making the plug useless. If the plaster crumbles, remove the screw and plaster the anchor in position, keeping the plaster out of the anchor.

PLASTER OVER BRICK

Only plugs will hold to plaster over brick walls. Buy a carbide-tipped masonry bit for drilling holes into the brick. You can use any plug-type anchor, lead plugs being the strongest. If you hit plaster over cement or stone, use plugs but with short screws. Keep the body diameter sufficient to fill the anchor.

BRICK, CONCRETE BLOCK, AND CEMENT

In these walls use a carbide-tipped masonry bit and drill for plugs. If you learn to use lead plugs properly, you'll find them reliable and effective.

PLASTER-AND-LATH OVER BRICK

There is one variation of the plaster-and-lath wall. It shows up among outside walls of apartments. Instead of two-by-four studs, the lath is supported by narrower pieces just ¾ by 1¼ inches and spaced as much as 18 inches apart. The ¾-inch space is too small for toggle bolts, the gap precludes the use of plugs, and the studs are nearly impossible to find. If you're hanging something really heavy, like a loftbed, look for the studs. For lesser loads, you can use a dowel. Purchase a ⅜-inch masonry bit and ⅜-inch doweling. Drill a hole through the plaster 1 inch into the brick. Clean out the hole and glue the dowel in with epoxy cement. Allow it to dry and drill for a screw or nail. If you are careful not to crumble the plaster, this is a very effective mounting.

SHEETROCK

In new and renovated apartments, Sheetrock-and-stud walls are the rule. To hang from these walls, three methods can be used. The first two, screws in studs and toggle bolts are the same as for a plaster-and-lath wall. (Incidentally, the stud finders you might find in hardware stores and lumberyards work only for Sheetrock walls.)

The third, developed especially for Sheetrock walls, is the molly bolt, which works by clamping the wall inside and out. Drill a hole for the molly, insert, and screw the bolt tight. This forces four wings to spread on the inside of the wall, squeezing the front and back nuts of the device together. When tight, unscrew the bolt and you have a threaded hole in the wall. There is a variation of the molly bolt for use in hollow-core doors.

WOOD PANELING

If your apartment has wood paneling, use the techniques of the plaster-and-lath wall. Check behind the paneling to see if it was put up over another kind of wall, in which case use the specific technique for that wall.

STONE

If you have stone walls (unusual in an apartment) try the carbide bit for lead plugs, and if that doesn't work seek professional help.

CEILINGS

Since anything attached to the ceiling will necessarily be suspended, special considerations are called for. Toggle bolts work best, especially for hanging plants. Finding the joists (studs which lie in the ceiling) is a must for heavy work like hanging chains for a loftbed. If the ceiling is so composed that these two methods are precluded, lead plugs or plastic anchors should do the job.

Plaster-and-Lath	*Solid Plaster*	*Plaster Over Brick*
screws in studs	plastic anchors	lead plug,
toggle bolts	fiber plugs	shortened plastic
	lead plugs	anchor with short screws

Plaster-and-Lath Over Brick	*Wood Paneling*
screws in studs	screws in studs
screw in dowel	toggle bolts

Brick, Concrete Block, Cement, or Stone	*Sheetrock*	*Ceilings*
lead plugs	molly bolt	toggle bolts
	screws in studs	screws in joists
	toggle bolts	lead plugs

X-BRACE

If you have plaster ceilings and wood floors, the x-brace can take the place of studs in the wall for many attachment applications, such as lofts, bookshelves, cabinets, plants. Using a scissors action, it wedges itself securely between ceiling and floor and can easily maintain a several-hundred-pound load. The band clamp is tightened by a wrench and held with a ratchet, so the snugness of the fit can be adjusted while the brace is in place. Also, the x-brace can be quickly knocked down and transported. If you're wary of putting holes in the wall, this device is for you.

CONSTRUCTION

Find the centers of the two-by-fours and in each cut a notch 1¾ inches by 4 inches. Place the notches together like a half-lap and drill for the carriage bolt, recessing the head.

Drill two ¾-inch holes about 18 inches from one end of each piece. Make the holes 3 or 4 inches apart and centered on the wide face of the two-by-four. Round the ends of the two-by-fours with a Surform plane.

Assemble the brace with carriage bolt and band clamp. To install, place one of the one-by-fours on the floor where you want studs. Open the x-brace so it fits between ceiling and floor. Stand it up on the floor one-by-four. Get on a ladder or stool and slip the ceiling one-by-four in position over the ends of the brace. The band clamp should be near the floor for easy access. (Alternatively, you can put it near the ceiling to keep it out of the way.) Tighten the clamp so the brace stands

Bill of Materials:

2 — 10′ two-by-fours (for 8′ to 10′ ceiling)

1 — ⅜″ carriage bolt with nut and washer

2 — one-by-four boards 3′ long

1 — 10′-to-12′ band clamp

by itself. The ends of the brace should be square on the one-by-fours. With the band clamp not too tight, you can adjust positions of pieces with hammer or mallet. When the brace is in final position, tighten the clamp so the band hums when plucked. Attach what you like with bolts or screws. To get an idea of the strength of the brace, attach a C clamp and hang from it, shaking from side to side. If there's any movement, tighten the band clamp.

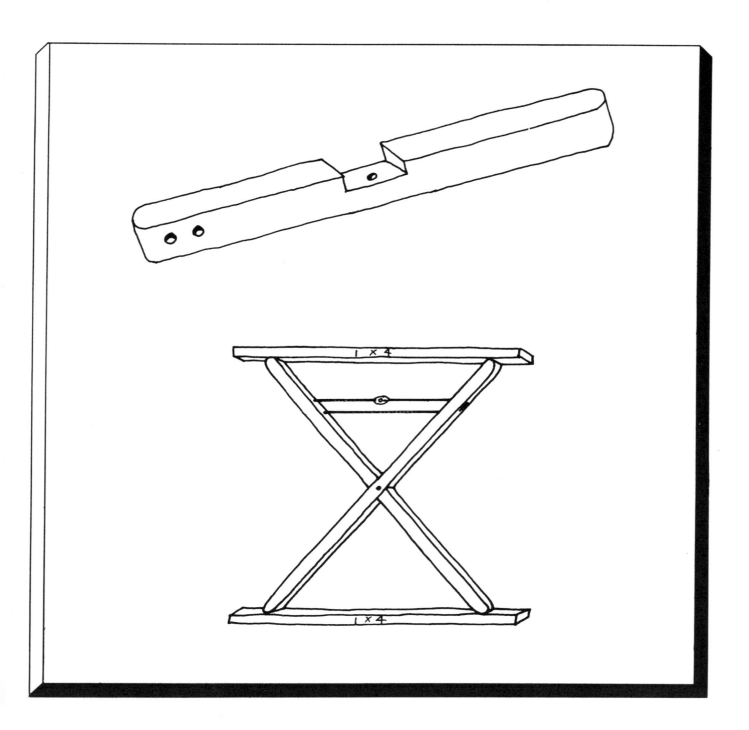

POLE BOOKCASE

This bookcase uses both a minimum of materials and the solidity of your floor and ceiling for a simple, solid structure capable of shelving an entire library. The shelves run from floor to ceiling, or anywhere in between, and can be placed anywhere in the room.

Two poles, 1½ inches in diameter, with one flat side, are fastened to the floor and ceiling. (Stair railings are manufactured with flat side.) This is done with the two-by-threes, which have holes to accept the poles and are attached to the ceiling with toggle bolts, to the floor with screws. The shelves are drilled so they can be strung on the poles, and the spacers are screwed to the flat side of the poles, supporting the shelves.

Drill the two-by-threes 3 inches in from their ends, centering the holes. Do the same for all the shelves. Draw a center line down one spacer of each kind, and drill two pilot screw holes on this line, 2 inches from each end. Use these spacers as templates and drill the rest of the spacers. Much of the drilling can be done with the pieces clamped together.

Drill the 3/16-inch holes for the toggle bolts in the two-by-threes. Thread all the shelves and the two-by-threes onto the poles and temporarily place into position. Drill the 5/8-inch holes in the ceiling to accept the toggle bolts, and attach the top two-by-three to the ceiling. Drill through the two-by-three sideways, for a ¼-inch dowel, and glue in a ¼-inch dowel 2½ inches long. Do this for both poles. The poles are now hanging from the ceiling, flat side in. Use a level to make them plumb (vertical) and attach the bottom two-by-three to the floor with nails or screws. You'll need someone to

Bill of Materials:

2 — stair railings, length equal to the floor-to-ceiling height, 1½″ in diameter

about 10–12 pairs of spacers to fit your books (8″, 10″, and 12″ are good sizes)

about 5–6 shelves, 3′ to 4′ (4′ for paperbacks)

2 — two-by-threes as long as the shelves

1½″ drill bit

1½″ screws, 2 to a spacer

4 — 3/16″ × 4″ toggle bolts

5/8″ bit for toggle bolts

hold all the shelves while you work near the floor. Attach the spacers to the insides of poles, starting from the bottom. A notched spacer may help at the floor, but is not necessary. Attach the spacers in pairs. You may want to leave a few shelves unattached at the top to leave room for expansion. This bookcase also makes an excellent room divider.

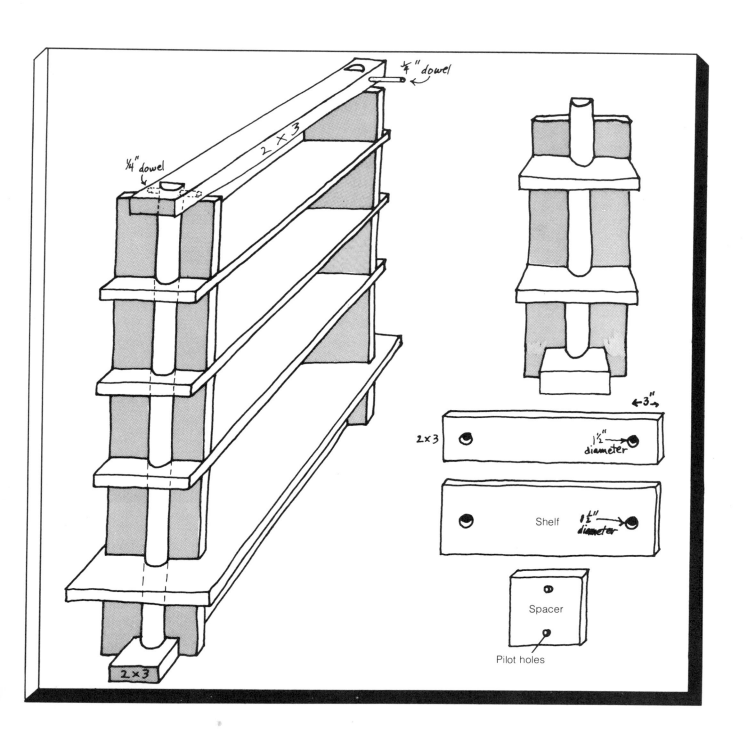

¼" dowel

¼" dowel

2 × 3

2×3

←3"→

2×3

1½" diameter

Shelf

1½" diameter

Spacer

Pilot holes

FREE-STANDING LOFT

With loftbeds, as with bridges, it's not the strength but the stability that counts. A sheet of plywood standing on four-by-four legs is certainly strong enough to support two people, mattress, and box spring. However, if the people like to roll over in their sleep, that sheet would sag and bounce like the skin of a waterbed. A structure to keep the plywood rigid is necessary. In addition, braces from that structure to the legs of the loft keep the leg joints from coming loose, increasing the stability still further. In lofts hung from the studs in the walls, the wall itself acts as a brace between the loft floor and the legs (studs). Most people building lofts like to attach at least one edge of the frame to the studs in the wall for stability. While this is recommended, the following loft is of free-standing design and can be placed anywhere in a room or dismantled and moved to a new apartment.

The following is a numbered list of jobs that must be done to build this loft.

1. Assemble materials, build notching template
2. Cut notches
3. Drill carriage-bolt holes
4. Assemble frame
5. Cut notches in plywood sheet
6. Attach sheet to frame
7. Cut braces
8. Drill braces for screws
9. Drill four-by-fours for carriage bolts
10. Attach legs to loft frame
11. Drill loft for brace screws
12. Attach braces to loft
13. Cut railings and attach to legs

1. Assemble materials, build notching template

After assembling all your materials begin construction by making a template for the twelve notches. The blade of the template can be a ⅛-inch or ¼-inch piece of hardboard accurately shaped so its width is exactly the thickness of the two-by-four you're using, and its depth half the width of the two-by-four. Let the fence of the

Bill of Materials:

4 — 8' four-by-fours	legs
1 — 4' × 8' × ¾" plywood	floor
1 — 2' × 8' × ¾" plywood	braces
3 — 4' two-by-fours	cross braces of frame
2 — 8' two-by-fours	long members of frame
1 — 8' two-by-four	railing
2 — 4' two-by-fours	railing

half pound 2" common nails

8 — ⅜" × 6" carriage bolts

washers, lock washers, and wrench to fit bolts

1 — 25/64" drill bit with ¼" shank

16 — 3" #10 flathead wood screws

48 — 1½" #10 flathead wood screws

6 — 4" #10 flathead wood screws

countersink

¾" spade bit

glue

Note: You may leave the narrow side open. For this use 2 – 8' two-by-fours and 1 – 4' two-by-four.

template project 3½ inches on either side of the blade for ease in marking the ten end notches (Figure 1).

2. Cut notches

With the notching template mark out the notches, then cut them. Check the fit as you go to insure ease of assembly later. Keep all the notches on the same side of the pieces, as in Figures 2, 3.

3. Drill carriage-bolt holes

Drill the holes for the carriage bolts with the 25/64″ bit. The holes through the long members should be drilled 1½ inches from the bottom edge (the edge without notches) and the same distance from the top edge of the cross braces (also the edge without notches). Use the notching jig to locate the holes. Keep the hole centered between the end of the piece and the notch. (You can use two four-by-fours as a drill table for these pieces if you have no table to work on. See Figure 7.)

4. Assemble frame

Assemble the frame with glue and 4-inch clamps. Spread the glue on the inside of the notches and directly above them (see Figure 4). Clamp the joints with moderate pressure, wiping the glue as it squeezes out from the joint. If well cut, the frame alone is incredibly strong—and even if it isn't, the floor will strengthen it. Set the frame aside and work on the plywood, or leave the frame on the floor and use it as a saw table for the plywood.

5. Cut notches in plywood sheet

The four corners of the plywood sheet should be notched out. This allows the four-by-four legs to rise above its surface and form the uprights for the railing. Use the adjustable square to mark 3½-inch squares in the corners of the sheet, and cut them. Since a full sheet of plywood is heavy and hard to move, plan how you will cut the corners. If you don't use the assembled frame as a saw table, lean the sheet against a wall and raise the corner you're working on with a four-by-four. Always have two people on hand when moving the sheet. See Figure 5.

6. Attach sheet to frame

When the notches are cut, you can attach the sheet to the frame. Clear the necessary floor space and place the frame flat on the floor. Spread a bead of glue along

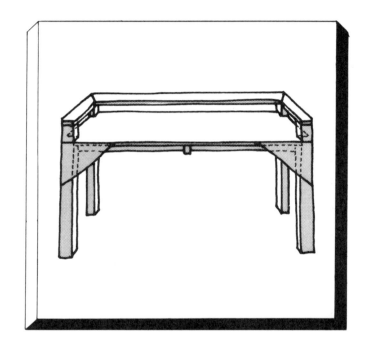

each top edge of the frame. With your helper, put the plywood sheet on the frame, good side up. The notches in the sheet should correspond to the open corners of the frame. Nail down the plywood with 2-inch nails, placing one every 6 inches along all the edges of the frame.

7. Cut braces

The braces are cut from the 2-by-8–foot sheet of plywood. First cut 2-by-2–foot squares, then cut these in half along a diagonal. For safety, cut off the tips of the braces ¾ inch back. If you want, you can personalize your loft by cutting a silhouette in the braces. To cut a silhouette, draw the design on one piece, drill the necessary clearance holes for the saber-saw blade, and cut out the design. After cleaning the rough edges with #60 sandpaper, use the pierced brace as a template for the rest. Remember that the braces will be in mirror-image

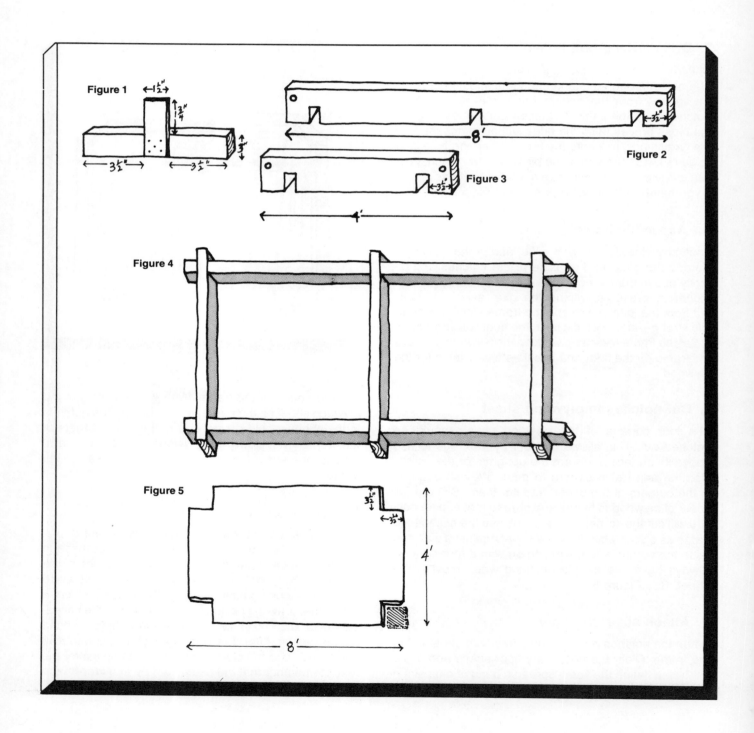

Figure 1

Figure 2

Figure 3

Figure 4

Figure 5

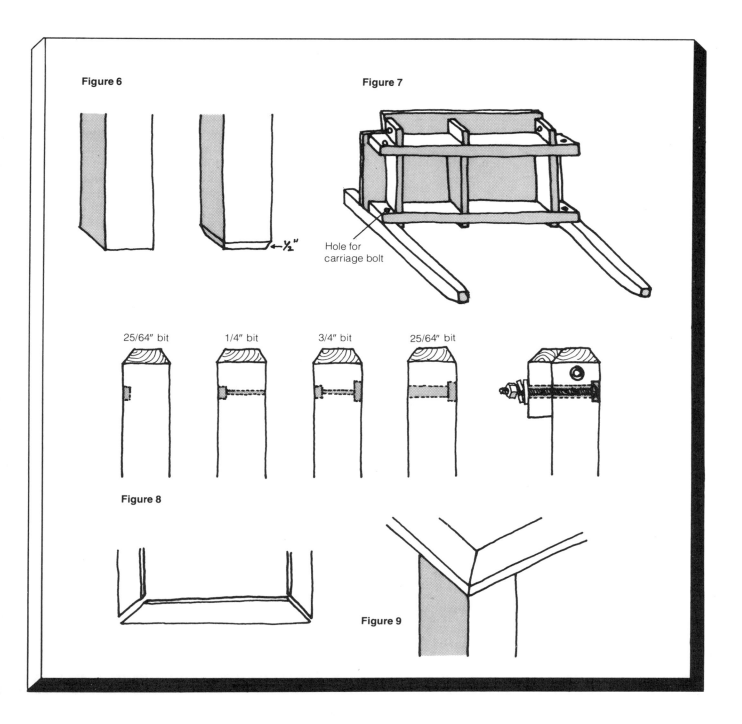

Figure 6

Figure 7

Hole for
carriage bolt

←½"

25/64" bit   1/4" bit   3/4" bit   25/64" bit

Figure 8

Figure 9

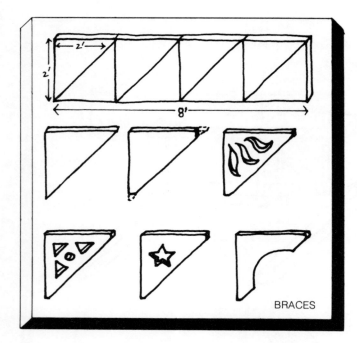

BRACES

pairs, with four lefts and four rights. Cutting silhouettes is a little bit of work, but the satisfactions can be great.

8. Drill braces for screws

To drill for screws, first draw a line ⅜ inch from the top edge of the braces, and another line 1½ inches in from the side. Drill three holes 6 inches apart along each of the lines with a 3/16-inch bit. After drilling the holes in all the braces, drill a conical hole with the countersink bit. This allows the screws to be driven flush with the surface of the brace.

9. Drill four-by-fours for carriage bolts

First plane off (bevel) the edges of the four-by-fours where they touch the floor (see Figure 6). Take off about ½ inch, so the feet won't splinter when you move the loft around. Now mark the height of the loft on the four-by-fours with a clearly visible line. Place two of the marked four-by-fours on the floor 8 feet apart and parallel. With an assistant, place the open corners of one edge of the frame on the four-by-fours (see Figure 7). Line up the top of the loft with the lines on the four-by-four legs. Drill from the inside with the 25/64-inch bit through the holes in the frame about ¼ inch into the four-by-fours (see Figure 8). Remove the legs, change to a ¼-inch bit, and drill the rest of the way through. Change to a ¾-inch spade bit and drill from the *outside* of the four-by-four about ½ inch in. This provides clearance for the head of the carriage bolt so the brace can be screwed over it. Now change back to a 25/64-inch bit and drill out the hole. Place some masking tape on each leg and label its position ("front left," for example). Make sure to label the loft also. Be sure to drill *both* holes in each leg. When two legs are done, remove them, flip the frame, and do the other two.

10. Attach legs to loft frame

After the second two legs are drilled, place them in position with the frame standing sideways as in the previous step. Slip a ⅜-inch carriage bolt through the four-by-four (it should slip in because a 25/64-inch hole gives 1/64 inch clearance for it) and into the frame. Slide on the washer, lock washer, and nut and tighten almost tight. When two legs have been so attached, flip the frame-legs assembly and do the second two legs. Note that only one bolt is in each leg at this point. When all four legs are bolted on, stand the loft upright, slip in the other bolts, and tighten up. Check the loft with the level: any wild irregularity indicates an incorrect height on one or more legs. These legs would then be removed and redrilled.

11. Drill loft for brace screws

Place the braces into the position and mark the screw holes on the legs and frame with an awl or nail. With an ⅛-inch bit, drill about ¾ inch in and attach the brace with screws rubbed on soap for lubrication.

12. Attach braces to loft

Attach one brace at a time, checking for the squareness

of each leg. You should feel the loft start to grow rigid at this stage, with less tendency to sway. When all the braces are attached, the structure of the loft is finished. It should feel solid and sturdy. If it rocks, place some shims made of thin wood or cardboard under the high leg (the rocking may be caused by a low spot in the floor). While the loft can be considered built at this stage, most people feel more comfortable on a loft with railings.

13. Cut railings and attach to legs

Cut 45-degree angles on one end of the 4-foot two-by-four railings and both ends of the 8-foot two-by-four piece and drill near the ends for long screws (see Figure 9). Be sure to countersink the holes if you're using flat-head screws. The three railing pieces form a squared frame that sits on top of the legs. Drill the ends of the legs for the screws, and attach the railings to the legs with 4-inch screws. One side stays open for the ladder access.

WALL-TO-WALL LOFT

A wall-to-wall loft can be most elegant. With no legs to interrupt the scheme, it looks suspended—a flying carpet hovering above the floor. Below, the beams create a beamed ceiling, which can be heightened with a contrasting stain. In this design, there are no interruptions of the wall-to-wall sweep of the beams, for a simple tasteful construction.

Begin by measuring the distance between the two walls the loft will span. Cut five of the two-by-fours to this length, fitting every one to the walls. Cut the sixth two-by-four in half, to 60 inches. (It is important that the two halves be the same length even if a shade under or over 5 feet.)

Beams

Cut notches in the ends of all the beams for a half-lap joint. Keep all the notches on the same side of the board. Mark and cut three more notches in each of the two cross pieces according to the dimensions in Figure 1. Use a cutout from an end notch to mark the width of the middle notches. Place the cutout to get the full width of the two-by-four. Use the adjustable square to mark the depth. Test all the notches for fit before screwing the crosspiece in place.

Floor of Loft

At first inspection, a 4-by-8 sheet of plywood just won't fit on a 5-by-8–foot loft without adding a piece. Unfortunately, the 1-by-8–foot piece added has only one two-by-four under it for support, which isn't enough. A novel solution to this problem is to cut the 4-by-8 sheet on a diagonal. Slide the two triangles to get a 5-foot width and cut off the tips (see Figure 2). Now the addition of a 2-by-5-foot piece completes the floor. In addition, all three pieces (the two trapezoids and the rectangle) are supported by several two-by-fours, insuring full support across the floor of the loft.

Assembly

With beams and floor cut, you're ready for assembly. Installation of a loft requires at least one assistant—

Bill of Materials:

6 – 10′ two-by-fours	beams	
1 – 4′ × 8′ × ¾″ plywood	top	
1 – 2′ × 5′ × ¾″ plywood	top	
80 - 6d common nails (2″)		
glue		

preferably several. The most help is needed at the very beginning, placing and installing the notched supports on the walls. First, position the notched support on one wall at the correct height and level (Figure 3). Draw a line along the bottom of the support, and set it aside. Locate the studs in the wall above this line. Attach the notched support to the wall with flathead screws, counter-sinking the holes for full support by the screws. If attaching the crosspiece to a masonry wall, use the appropriate method.

To locate the second crosspiece, place two beams in their respective notches in both the attached and the floating support. Bring the floating crosspiece to the wall, positioning it so that the piece is level and the beams are both level and square with the crosspieces. This is the operation that takes troops. If you are building alone, position the second crosspiece by measuring from the third wall and the floor. With one screw in, check with a beam and level for squareness and levelness. If true, put the rest of the screws in, one for each stud or one every 15 inches. With the crosspieces up, you can set in the beams. If one of the beams runs along a wall, attach it to that wall for greater strength. When all the beams are up, you can attach the floor.

Nail the three pieces of flooring (the two trapezoids and the rectangle) to the beams of the loft with 2-inch nails (Figure 4). Trim any projecting edges, if necessary, with the saber saw. Attach a ladder and, if you feel uncomfortable without a railing, see the railing for the free-standing loft.

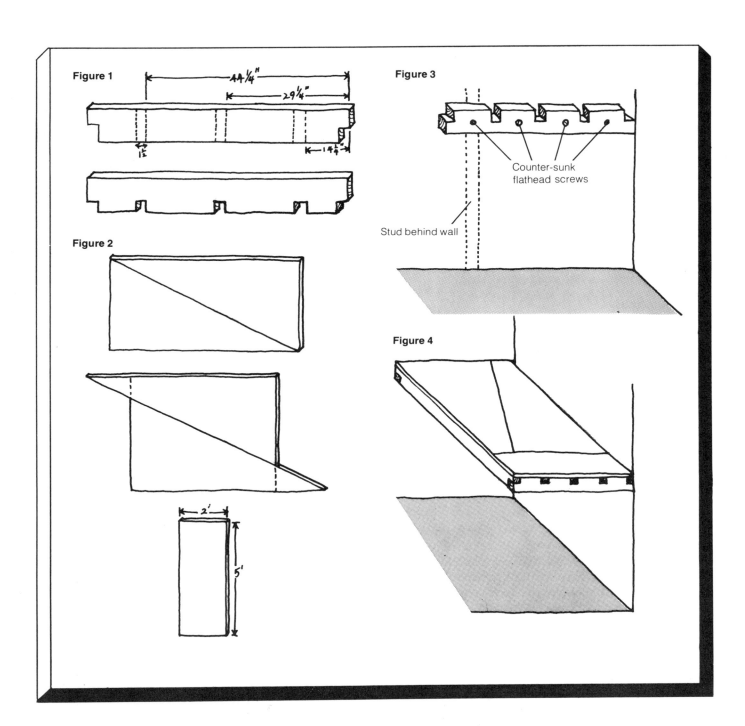

Figure 1

44 1/4"

29 1/4"

1 1/2"

14 1/4"

Figure 2

2'

5'

Figure 3

Counter-sunk
flathead screws

Stud behind wall

Figure 4

ROOM DIVIDER

Begin by making a notch template 5⅝ inches long for the thirty notches. Mark the notches, leaving 12 inches between each, as shown in Figure 1. Cut all the notches. Here a ¾-inch wood chisel can come in handy. Simply cut in the long sides of the notch, place the edge of the chisel on the line, bevel facing toward the notch, and tap with a hammer or a mallet. The waste will drop out. After all the notches are cut, assemble the shelves and uprights. Prenail all the squares, putting one nail in each corner about ⅜ inch in. Try to have extra squares on hand in case some split. Glue and nail all the squares to the crossings. Place the divider in position permanently by screwing the one-by-twos to the floor and uprights, as shown in Figure 2. This technique of case construction can be adapted to any size. For a different look try discs at the crossings instead of squares.

Bill of Materials:

3 — 75¾″ one-by-twelves

5 — 50¼″ one-by-twelves

30 — ½″ × 2″ × 2″

3 — 11¼″ one-by-twos

70 finishing nails (1½″)

optional, ¾″ wood chisel

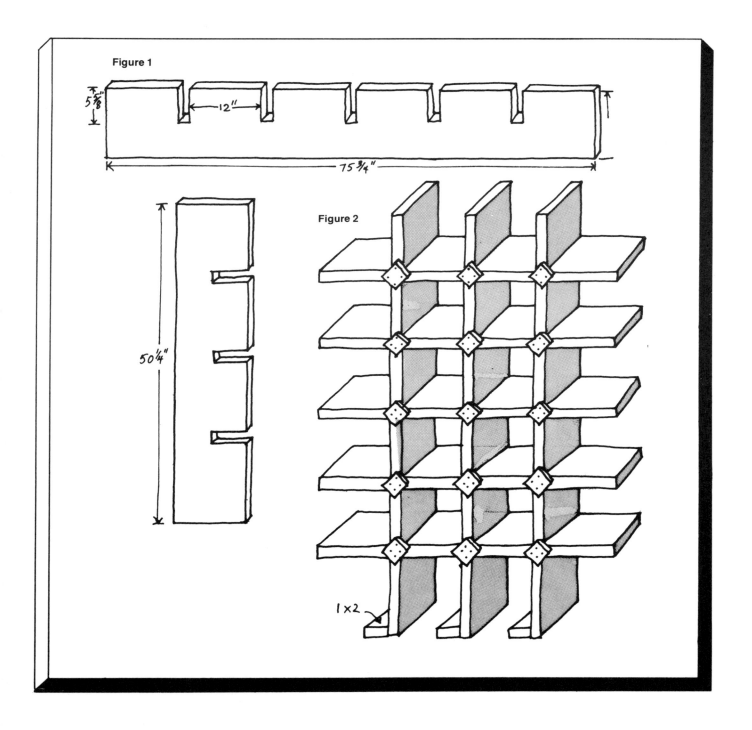

Figure 1

5⅝"

12"

75¾"

50¼"

Figure 2

1×2

While it would be somewhat cheaper to build one large bookcase for a wall, building several units offers the possibility of portability for a future move. In addition, large projects are hard to handle during assembly. The following three basic units—half-lap with back brace, rail-spacer, and double-notched—all result in the same exterior dimensions but demonstrate three different methods of construction. The first two have attachments that can increase the utility of the total unit.

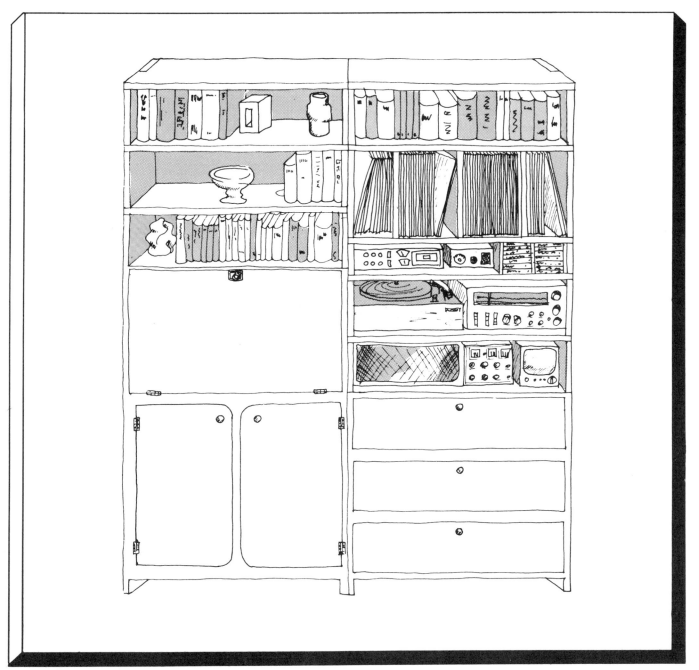

HALF-LAP WITH BACK BRACE

This wall unit is simple to construct and yet quite strong. Using a series of half-lap joints, the shelves are set into the uprights and the case is then squared and secured with a back.

Begin by constructing a notch template for marking the thirty-two notches. Mark the sides and shelves as shown in Figure 1 and cut the notches. The uprights can be cut together. Check your notches for fit before assembly. To construct, spread glue where necessary, fit all the shelves to one upright, and nail them all in. Flip this assembly around so the attached upright is on the floor, and fit the other upright to the shelves. Nail on the second upright. Rotate the case face down (the notches of the upright open toward the face) and nail the back onto one of the uprights. Square the case with the carpenter's square and attach the back to the other upright and all the shelves.

Bill of Materials:

8 — 24″ one-by-ten pine shelves

2 — 84″ one-by-ten pine uprights

1 — ¼″ × 24″ × 84″ plywood back

4d finishing nails (1½″)

glue

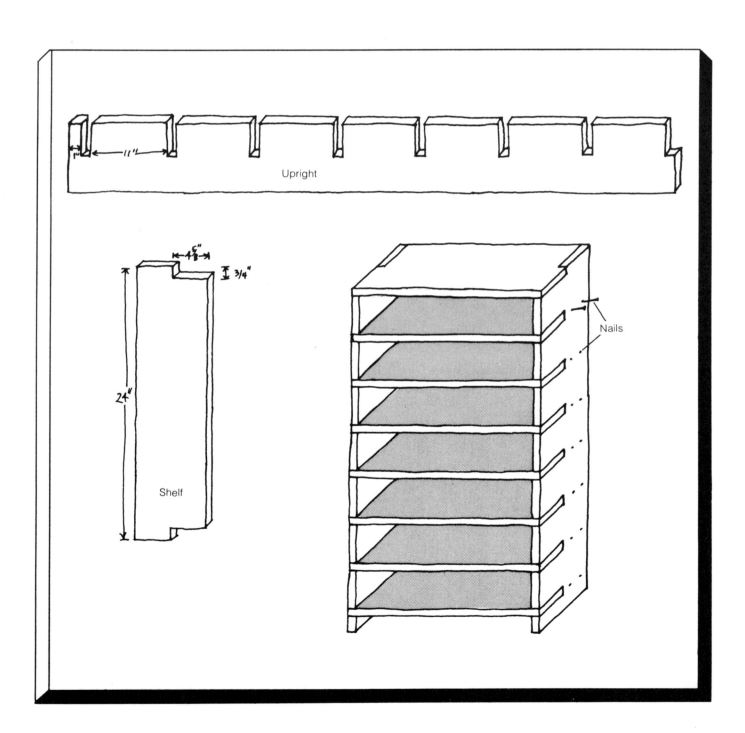

Upright

3/4"

1 5/8"

24"

Shelf

Nails

Some possessions are best stored behind closed doors. Achieving parallelism in hinges can be a real trial and often requires repeated attempts before it is reached. Here is a simpler method for hanging doors which can be added to any storage unit.

Draw the outline for the two doors with ruler and compass as detailed in Figure 1. Make a pocket cut (a cut started in the middle of a board) along the top or bottom edge of each door and cut halfway around, on the hinge side (Figure 2). Attach the piano hinge across the cut line with the ridge of the hinge directly over the cut. Do this for both doors. Now continue the cuts around the doors so they are released from the board. Attach the door unit across two shelves and two uprights of the wall unit with glue and nails or screws. Knobs and catches can be added if desired. Surform and sand the edges of the doors for smooth operation.

Bill of Materials:

1 — ¾" × 12½" × 24"

2 — 10" piano hinges with ¾" screws

2 knobs, 2 catches

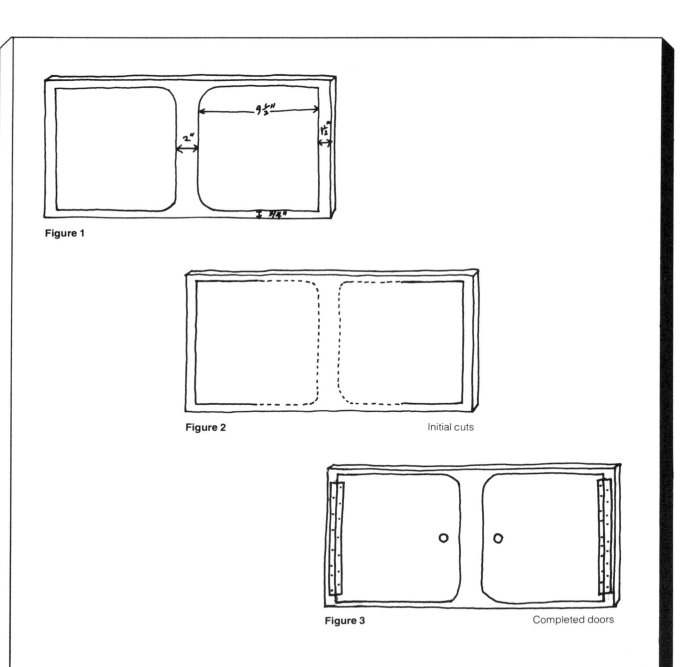

Figure 1

Figure 2 Initial cuts

Figure 3 Completed doors

Begin by marking and cutting a ¾-by-¾-inch notch in the four corners of each shelf. Glue and nail each shelf to a pair of spacers (see Figure 2). Now put two rails on the floor 22½ inches apart and place two sets of shelf-spacer assemblies on them. The 3-inch spacer should be at the bottom, or end, of the rails. Clamp the rails to the spacers so they don't slip out when you're nailing. Starting from the bottom, glue and nail the front rails to the spacers and shelves. With the bottom unit nailed in, butt up the next shelf-spacer unit, add a little glue to the joint, and nail in the front rails. Repeat all the way along the rails. Now flip the unit and nail in the back rails. With all four rails connected, nail on the back. The spacer dimensions are merely suggestions and are easy to change. Just remember the sum of the thickness of the shelves and the height of the spacers must equal the length of the rails.

Bill of Materials:

8 — 24″ one-by-ten pine shelves

10 — ¾″ × 7¾″ × 11″ ⎫ spacers

12 — ¾″ × 7¾″ × 10″ ⎬ one-by-ten cut

4 — ¾″ × 7¾″ × 3″ ⎭ 1½″ narrower

4 — ¾″ × ¾″ × 83″

1 — ¼″ × 24″ × 83″

4d finishing nails (1½″)

glue

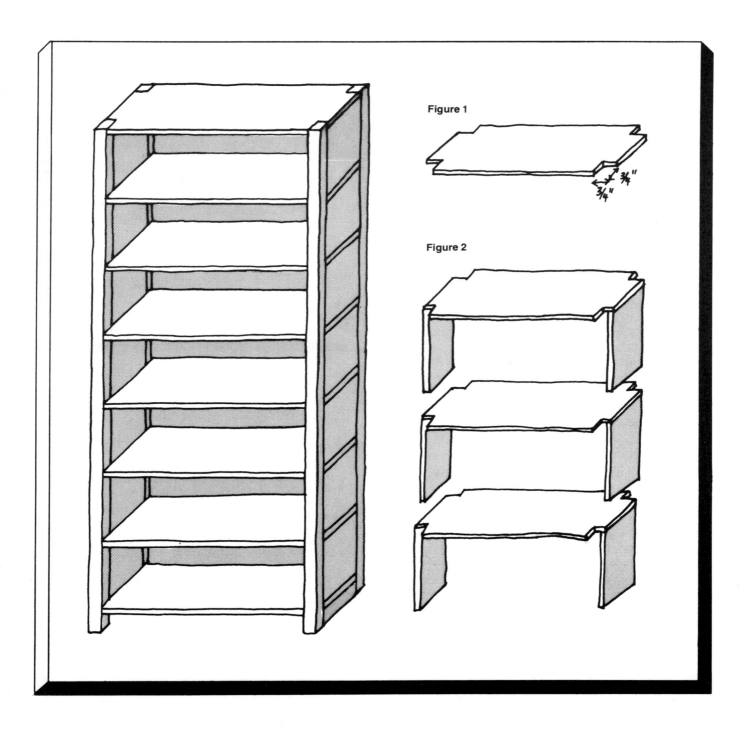

Figure 1

3/4"
3/4"

Figure 2

Attachment: Drawers

Drawers are often useful. This set is self-contained and can be attached between two shelves of a bookcase, under a desk top, on a table, or loose.

It's best to build the case first because you can fit it to the shelves and then fit the drawers to the case. The case is built using the false-dado technique, which lends itself to small construction.

Using a scrap piece of ¼-inch plywood as a slot template, glue the spacers to the sides with clamps (1-inch clamps are nice here) or by nailing (Figure 1). If you nail, use four-penny nails and nail both boards into a third, leaving the heads exposed for later removal (Figure 2). With all the spacers in, glue and nail in the shelves, top, and bottom with ¾-inch nails. The shelves should be flush with the front of the case. Attach the back, which should fit between the sides, with glue and ¾-inch nails. The case is now complete, and can be fitted into place using glue, nails, or screws (Figure 3).

There are three drawers to this case, two 3 inches deep and one 4 inches deep. All are large enough to hold typing paper or standard-size magazines. Concerning the materials: ⅛-inch plywood can often be found in discarded drawer bottoms; ¼-inch plywood occurs as panels or backs of cabinets.

Note in the Bill of Materials there are two sets of fronts, one ⅛-inch and one ¾-inch. These drawers are first built as boxes using the ⅛-inch pieces front and back. After the box is complete the ¾-inch fronts are glued on. This allows you to adjust the drawer fronts so they come out neatly aligned.

Construction is straightforward. Glue and nail the bottoms each to a pair of sides. Nail and glue the fronts and backs to their respective side-bottom assemblies (see Figure 4). Tap a few ¾-inch nails through the ⅛-inch front of the drawers near the sides, so their tips are just exposed. Slip the drawers into the case, and press the pine or ply fronts onto the drawers so the nails dimple the wood. Remove the drawers, glue on the fronts, and nail together. For a simple handle, glue on a ¾-by-¾-by-4–inch strip of wood, centered on the face of the drawer. There should be enough clearance, but if the drawers stick, use soap, silicone spray, or sandpaper.

Bill of Materials:

Case

2 — ¼″ × 9¼″ × 11″   sides
4 — ¼″ × 3″ × 9″   spacers
2 — ¼″ × 4″ × 9″   spacers
2 — ¼″ × 9¾″ × 14½″   top, bottom
2 — ¼″ × 9″ × 14½″   shelves
1 — ¼″ × 11″ × 14½″   back

Drawers

2 — ¾″ × 3″ × 15″   fronts
1 — ¾″ × 4″ × 15″   front
4 — ¾″ × 8⅝″ × 2⅝″   sides
2 — ¾″ × 8⅝″ × 3⅝″   sides
3 — ¼″ × 8⅝″ × 3⅝″   bottoms
4 — ⅛″ × 2⅞″ × 13⅞″   fronts, backs 3″ drawer
2 — ⅛″ × 3⅞″ × 13⅞″   front, back 4″ drawer

one pound 4d finishing nails (1½″)

glue

four ounces ¾″ nails

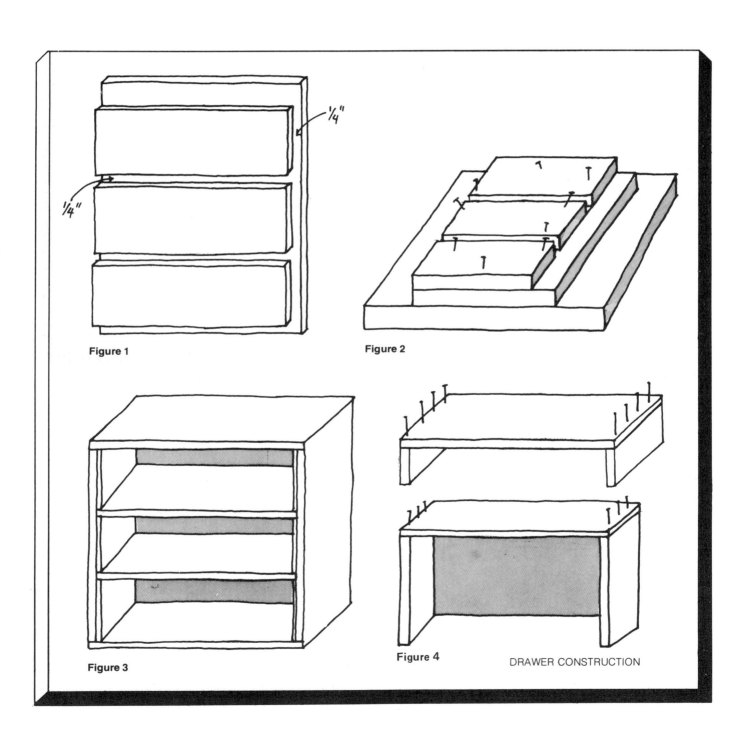

Figure 1

¼"

¼"

Figure 2

Figure 3

Figure 4

DRAWER CONSTRUCTION

Begin by marking the ¾-inch notches on the uprights as detailed in Figure 1. Clamp the two uprights together and cut the notches. Mark and cut the notches of the shelves as shown in Figure 2, and test each one for fit before assembly. Glue and nail the uprights to the shelves and the shelves to the uprights (two-direction nailing) (see Figure 3). Nail on the back to complete the structure.

To connect the units, screw them together using 1¼-inch flathead screws. Use at least six for each pair of units, two each at the top, the bottom, and near the middle. Be sure to countersink the screws so they don't show.

Bill of Materials:

8 — 24″ one-by-tens	shelves
2 — 83″ one-by-tens	uprights
1 — ¼″ × 24″ × 83″ plywood	back
4d finishing nails (1½″)	
glue	

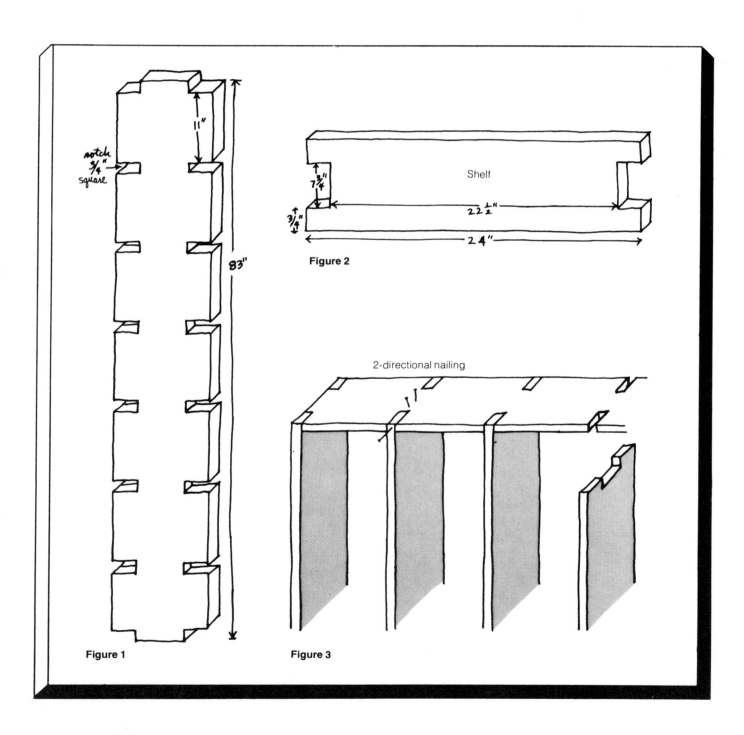

11"

notch
3/4"
square

83"

Figure 1

Shelf

7 3/4"

3/4"

22 1/2"

24"

Figure 2

2-directional nailing

Figure 3

ANGLES

(constructed with carpenter's square)

This table of angles includes all commonly used angles and all the angles used in this book.

Angle	Long Arm	Short Arm
15°	17¼"	4⅝"
22½°	16"	6⅝"
30°	12⅛"	7"
36°	16"	11⅝"
45°	14"	14"
60°	7"	12⅛"
72°	6½"	20"
75°	4⅝"	17¼"

DRILL SIZES FOR FLATHEAD WOOD SCREWS

Screw Size	Body Hole Size	Pilot Hole Size Soft Woods
0	1/16"	not used
1	5/64"	1/32"
2	3/32"	1/32"
3	7/64"	3/64"
4	7/64"	3/64"
5	1/8"	1/16"
6	9/64"	1/16"
7	5/32"	1/16"
8	11/64"	5/64"
9	3/16"	5/64"
10	3/16"	3/32"
11	13/64"	3/32"
12	7/32"	7/64"
14	1/4"	7/64"
16	17/64"	9/64"

Use pilot hole 1/32" larger for hardwoods (oak, walnut, maple, birch).

NAIL SIZES

Pennies	Inches
2d	1″
3d	1¼″
4d	1½″
6d	2″
8d	2½″
10d	3″
16d	3½″
20d	4″

LUMBER SIZES

Purchasing Name	Actual Dimensions
one-by-one (1″ × 1″)	¾″ × ¾″
one-by-two (1″ × 2″)	¾″ × 1½″
one-by-three (1″ × 3″)	¾″ × 2½″
one-by-four (1″ × 4″)	¾″ × 3½″
one-by-six (1″ × 6″)	¾″ × 5½″
one-by-eight (1″ × 8″)	¾″ × 7½″
one-by-ten (1″ × 10″)	¾″ × 9¼″
one-by-twelve (1″ × 12″)	¾″ × 11¼″
one-by-fourteen (1″ × 14″)	¾″ × 13¼″
two-by-two (2″ × 2″)	1½″ × 1½″
two-by-three (2″ × 3″)	1½″ × 2½″
two-by-four (2″ × 4″)	1½″ × 3½″
four-by-four (4″ × 4″)	3½″ × 3½″

LUMBER SIZES

1 × 1

1 × 2

1 × 3

1 × 4

1 × 6

1 × 8

1 × 10

1 × 12

1 × 14

2 × 2

2 × 3

2 × 4

4 × 4

BIBLIOGRAPHY

The Complete Book of Furniture Repair and Refinishing, by Ralph Parsons Kinney. New York: Charles Scribner's Sons, 1950.

The best of the many books on refinishing. Follow Kinney faithfully. As he says in the preface, "Anyone can turn out creditable work if he or she will but sincerely try."

DeCristoforo's Complete Book of Power Tools, by R. J. DeCristoforo. New York: Times Mirror Magazine, Inc., 1972.

DeCristoforo's masterpiece. Now an editor at *Popular Science Monthly,* R. J. DeCristoforo has written do-it-yourself articles on woodworking for many years. Specializing in simple jigs that solve difficult problems, DeCristoforo has assembled nearly all his tricks under one cover. Apartment carpenters should take special notice of the chapters on sabre saws and electric drills.

The Illustrated Encyclopedia of Carpentry and Woodworking Tools, Terms and Materials, by Stanley Schuler. Chester, CT: Pequot Press, 1973.

A useful reference for quick definitions.

A Museum of Early American Tools, by Eric Sloane. New York: Ballantine Books, 1964.

The dedication says it: ". . . these ancient implements are symbols of a sincerity, an integrity, and an excellency that the craftsman of today might do well to emulate." Besides that, the drawings are a delight, and the text richly informative, making *Museum* a must for any carpenter-craftsman's bookshelf.

The Nature and Art of Workmanship, by David Pye. New York: Van Nostrand Reinhold, 1971.

An essay on the difference between things made by hand and made by machine, with insights important to any craftsman or consumer.

Nomadic Furniture, by James Hennessey and Victor Papanek. New York: Pantheon Books, 1973.

Nomadic Furniture is both a catalog of available designs and a compendium of new ideas for the apartment dweller. Simply put, it is "How to build and where to buy lightweight furniture that folds, inflates, knocks down, stacks, or is disposable and can be recycled." A must.

Nomadic Furniture Two.

More good stuff from the folks who brought you *Nomadic Furniture.* If you like one you'll want both.

The Pine Furniture of Early New England, by Russell Hawes Kettell. New York: Dover Publications, 1929.

This is a classic. With over two hundred examples of all types of furniture, it encompasses a 300-year tradition for you to draw on, design from, or simply admire.

Public Works: A Handbook for Self-Reliant Living, compiled and edited by Walter Szykitka. New York: Links Books, 1974.

This contains a good section on tools and construction, which would be of particular interest, and a section on everything else from fire making to how to drive (random sampling), of general interest. This is an excellent source book for every home.

Reader's Digest Complete Do-It-Yourself Manual, Reader's Digest Association. New York: W. W. Norton & Co., 1973.

While designed for homeowners, this book is an indispensable reference for any work done in an apartment, from pounding a nail to four-wall renovation. It includes a projects section, which is indexed to the individual skills needed from earlier sections. The Manual is indeed complete, and should be the first place to look for any household technology question.